T0212407

Fundamentals of Human Physiology

Giovanni Cavagna

Fundamentals of Human Physiology

 Springer

Giovanni Cavagna
Milano State University
Milan, Italy

ISBN 978-3-030-19406-2   ISBN 978-3-030-19404-8   (eBook)
https://doi.org/10.1007/978-3-030-19404-8

This Springer imprint is published by the registered company Springer Nature Switzerland AG
The registered company address is: Gewerbestrasse 11, 6330 Cham, Switzerland

To my students

Foreword

In the fall of 1989, as a third-year medical student, I was preparing to enroll in a human physiology course at the University of Milan. Several professors offered that course and a lottery mechanism matched students with a professor. If your grades were high enough, though, and you were not happy with the results of the lottery, you had the option to petition for enrollment with a professor who had a reputation for being quite lenient. When the lottery assigned me to Prof. Cavagna, known for being particularly demanding, I heeded peer pressure and petitioned, and obtained, to be switched to the more lenient professor's course. Nonetheless, I remained intrigued by the exacting reputation that surrounded Cavagna, so I decided that, for the first week, I would attend classes of both professors. Attending those first few classes from Cavagna convinced me that his (deserved) reputation was a reflection of the quality, dedication, and commitment to teaching, and I re-petitioned to be switched back to where destiny had originally placed me. While, admittedly, I did not have the time to study any subject other than physiology for the entire academic year, I now look back at that course 30 years later and recognize that it was one of the best professional investments for my career. It is thus particularly precious that Cavagna has collected and structured the teaching material from 40-plus years of didactics to medical students in the current volume.

When approaching this book, one recognizes fundamental themes that transcend individual topics and chapters because they form the conceptual underpinning of physiology as one of the foundational disciplines of medicine: the concept of functions of organs and systems as transformations of energy, the quantitative approach to measuring such functions, the energy economy of such transformations, the integration of the components of a system across scales, from micro to macro, and the concept of homeostasis are all examples of those themes.

Chapter 1 deals with the statics and dynamics of the systemic and pulmonary circulations. Several physiologic concepts are introduced in the context of their clinical correlate, which will make the reading of this volume enticing for medical students. For example, the hydrostatic indifferent point is discussed in the context of the genesis of orthopnea in congestive heart failure, laminar and turbulent flow in relation to murmurs, the sphygmic waveform in relation to aortic valve disease, and the critical closing pressure in the setting of reduced flow states such as hypov-olemic shock and atherosclerosis. In addition, the rationale for proper clinical measurement of physiological variables, such as invasive blood pressure monitor-ing with intravascular catheters, is presented, and common missteps are discussed.

Chapter 2 treats the physiology of muscle, both skeletal and cardiac, and locomotion. While the heart is more commonly treated in conjunction with the circulatory system, the organization that Cavagna provides enables an immediate and informative parallelism of similarities and differences between the two types of striated muscle. This is the chapter in which the integration of molecular and systems physiology is most apparent, reflecting Cavagna's decades of seminal contributions to the physiology of the musculoskeletal system, from isolated muscle fibers to elephants.

Chapter 3 pertains to the ventilatory function and metabolism. Having pursued a career in the field of anesthesiology and critical care, it is enlightening for me to see how concepts such as the alveolar opening pressure during re-expansion from a collapsed state and the relaxation curve of the respiratory system have represented the conceptual foundation for the improvement in mechanical ventilator manage-ment of patients with acute respiratory distress syndrome that has led to improved outcomes over the past decade. The next frontier in this field is the development of personalized mechanical ventilation through, among other means, measurement of esophageal pressure to differentiate the relative contributions of the lung and chest wall to the impairment of respiratory mechanics in the individual patient. Consequently, the discussion of esophageal pressure measurements to tease apart the lung and chest wall curves is particularly timely. In addition, topics such as acid–base equilibrium and related disturbances are of the utmost clinical relevance as virtually any physician will have to interpret a patient's blood gas report at some point in their career.

Chapter 4 is on renal physiology, which is treated with an emphasis on the same overarching themes of the entire book: work (osmotic in the case of the kidney), homeostasis, and the functional unit (the nephron in the kidney). In addition, concepts that clinicians will encounter daily in their profession, such as the renal clearance of solutes, are explained in detail.

Five years after my physiology course, I went back to Prof. Cavagna and joined his laboratory as a part-time research fellow for a couple of years, before emigrating to the USA. I thus had the opportunity to learn from Cavagna the researcher, rather than the professor. The obsessive compulsive nature of his data checking, the attention to detail in experimental setup and conditions, and the importance of always asking "why" (i.e., intellectual curiosity) and of stripping away the irrele-vant to get to fundamental principles for how things work (i.e., theory) are all things that I learned in those later years. These teachings have been invaluable in my career and clearly transpire from this book.

Saint Louis, MO, USA   Guido Musch

Preface

In my frequent visits to Milan hospitals, I met several old students of mine who warmly told me that my lessons are still useful for their work. Since my last lessons on circulation, muscle, respiration, and kidney were filmed, I decided to publish them in the hope to help a larger number of students.

In my opinion, a physiological mechanism can be fully understood and more easily retained when it is developed from and through physics (when possible). This is what I did in my lessons assuming an elementary knowledge of mathematics and physics.

A problem in teaching is the separation between lessons content in student memory and his outside world experience. Often, the different topics are stored sequentially in the memory of the student, as in closed wagons of a train, to be "expectorated" at the examination (first figure below). This leads to two problems: (i) It becomes difficult for the student to explain an outside condition on the basis of the mechanisms learned during the course, and (ii) it becomes difficult to connect between them arguments taught in subsequent order. To prevent these inconveniences, it is necessary to open the doors of the train of student memory toward its daily external world and to associate items taught in a different order (second schema below). This is a task of both teacher and student. I did my best here.

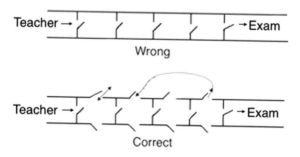

I reproduced the slides and my drawings on the blackboard as figures in the text. In the legend of the figure taken from several books and articles, I referenced the original source for further reading.

I wish to thank Dr. Mario Legramandi for his help in recovering original literature sources and his accurate revision of the manuscript.

Milan, Italy   Giovanni Cavagna

Contents

Chapter 1
Circulation of Blood

Abstract Definition of absolute, relative and transmural pressures. Manometers. Mechanical energy of a fluid. Statics of circulation: effect of height, acceleration and immersion in water on blood pressure and distribution within the body. Dynamics of circulation. Laminar motion: derivation of the parabolic profile of blood velocity within a vessel, the law of Poiseuille and its consequences. Axial accumulation of red cells: effect of vessel dimensions and the Fahraeus-Lindqvist effect. Flowmeter. Turbulent motion: derivation of the critical velocity and of the Reynolds Number. Incidence of kinetic energy in circulation: downstream pressure, side pressure and end pressure in different parts of the circulation. The sphygmic wave. Measurement of blood pressure. Organization of systemic and pulmonary circulation. Equilibrium conditions of blood vessels: definition of tension and the Law of Laplace. Elastic, active and mixed tensions in vessels walls. Critical closing pressure. The capillaries. Function of capillaries. Factors affecting production of the interstitial fluid and edema. Venous circle. Pulmonary circle from top to bottom of lung zones in the erect position: waterfall effect. Coronary flow: effect of heart beats frequency. Fetal circle.

1.1 Introduction

Physiology: a study of the functions of a living organism. What is a 'function'? A function consists in a transformation of energy. All the functions of our organism consist in a transformation of chemical energy into another form of energy. For example in the nervous system chemical energy is transformed in a difference of electrical potential across the neurons membrane (the action potential), in the muscular system chemical energy is transformed in a very different kind of energy: the mechanical energy (mechanical work done by muscle), the kidney transforms chemical energy in a difference of osmotic potential (when concentrating urine), the liver and other tissues transform chemical energy into different forms of chemical energy. A 'motor' is a transformer of energy: a living organism is an ensemble of small motors. These motors, in order to continue their function, require a *constant surrounding*. For example, they cannot work properly when the temperature increases above 45 °C, when

© Springer Nature Switzerland AG 2019
G. Cavagna, *Fundamentals of Human Physiology*,
https://doi.org/10.1007/978-3-030-19404-8_1

osmolarity deviates appreciably from 0.3, the pH from 7 or toxic molecules are produced by the metabolism. In conclusion, a constant internal surrounding is necessary.

The problem is that a motor pollutes its surrounding. For example consider the oxidation of a mole of glucose:

$$C_6H_{12}O_6 + 6O_2 \rightarrow 6CO_2 + 6H_2O + 686 \text{ kcal}$$

It requires oxygen and produces carbon dioxide, water and heat, i.e. per se the reaction implies a modification of its surrounding. Then what?

If you want to maintain constant an inanimate object, for example a picture, you put it in a closed container. On the contrary if you want to maintain constant the surrounding within the living body, you, cannot close its communication with the external ambient. In fact, as shown in the example above, a closed system would be polluted by an increase in carbon dioxide, water and heat. In conclusion: a function requires a constant internal surrounding, but when it occurs, it pollutes it. The solution of this problem is to open the system towards the external environment. For example, in the example above, the function to go on must absorb oxygen from outside, and eliminate carbon dioxide, water and heat into the external surrounding. In other words in order to perform work a function requires both an input and an output. This means that all the cells functioning in our body, for example in the liver, must communicate with the external environment in spite of the distance between internal and external ambient. How this is done?

Two ways: one way appropriate for large distances, the second for small distances. The first one is the *circulation* of the blood. In the example above, the blood of the capillaries near the cells releases oxygen and takes away carbon dioxide, water and heat to large distances. However as in a bus network, the bus does not arrive within your home and you must walk from the bus stop to your apartment by walking, i.e. from the capillary to the working cell. In our body system this walking mechanism corresponds to *diffusion*, a mechanism useful for small distances only, operating in all tissues of our body to allow communication of the internal surrounding with the external environment.

The entire circulation of blood is at the service of the capillaries: blood flows one-way only contrary to respiration where air flows two-ways. As you know from anatomy blood flows from the left ventricle, into aorta, arteries, arterioles, capillaries, cava veins, right atrium (systemic circulation), right ventricle, pulmonary artery, pulmonary capillaries, pulmonary veins, left atrium (pulmonary circulation) (Fig. 1.1).

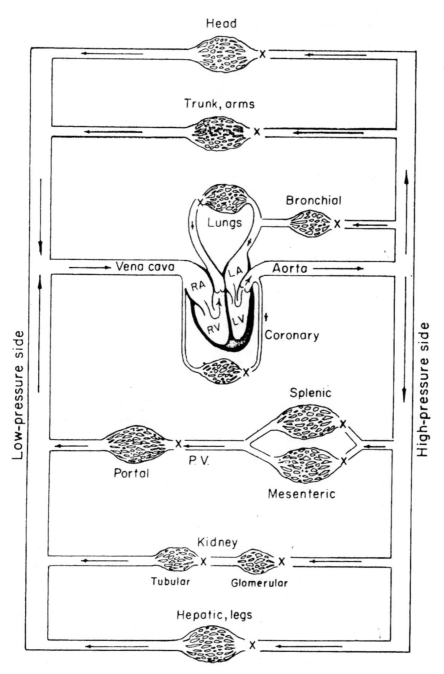

Fig. 1.1 The ×'s indicate the location of control points where arterioles may control the flow. *R.A.*, right atrium; *L.A.* left atrium; *R.V.*, right ventricle; *L.V.* left ventricle; *PV* portal vein (from Green HD in Glasser O [ed]: *Medical Physics*. The Year Book Publishers, Inc., 1949)

1.2   Blood Pressure and Mechanical Energy

The pressure of blood differs from the pressure outside the body for two reasons: the gravity (statics) and the active contraction of heart ventricles (dynamics).

It is important at this point to understand the difference between *relative* and *absolute* pressures:

$$P_{relative} = P_{absolute} - Pb \text{ (barometric pressure, e.g.760 mm Hg)}$$

In general in physiology and medicine, one always refers to the *relative* pressure. For example a pressure of 120 mm Hg means

$$120 \text{ mm Hg} = 880 \text{ mm Hg} - 760 \text{ mm Hg}$$

On a mountain, in the same physiological conditions:

$$120 \text{ mm Hg} = 620 \text{ mm Hg} - 500 \text{ mm Hg}$$

Even more important is the *transmural* pressure:

$$P_{transmural} = P_{absolute} - P_{outside \ the \ vessel}$$

The transmural pressure equals the relative pressure when the pressure outside the vessel equals the atmospheric pressure, as when the tissues between vessels and surface of the body are very slack. In other cases, as in the eye where the pressure is about 20 mm Hg, the transmural pressure is lower than the relative pressure, i.e. the retinal vessels will have a greater tendency to collapse.

In conclusion: when the pressure outside our body decreases the pressure applied to the wall of the vessels, i.e. the gradient of pressure between inside and outside of the vessel, *remains the same*. This is in contrast with an air filled balloon where the gradient of pressure between inside and outside increases with height. In fact the pressure inside the blood vessel is due to the active contraction of heart ventricles (dynamic) and the gravity (static), whereas in the gas of the balloon pressure depends, as we will see, from the kinetic energy of the gas molecules $\left(P = nRT / V \right)$. Only if the pressure outside the body falls below its vapor pressure, the blood would 'boil' and the vessels would tend to explode.

In what follows we will always refer to the *relative* pressure. This pressure falls from an average of about 100 mm of mercury (Hg) in the aorta down to about 2 mm of Hg in the cava veins, i.e. $100 - 2 = 98$ mm Hg is the gradient of pressure that sustain motion in the systemic circulation; in the pulmonary circulation the gradient of pressure from pulmonary artery to veins is much smaller, about 15 mm Hg.

What promotes the motion of a fluid (the blood in our case)? From what stated above one would infer that the motion is sustained by a difference in pressure. The pressure is a force F acting perpendicularly on the unit of a surface S, i.e.

$$P = F/S$$

And dimensionally, since $F = mlt^{-2}$, i.e. the mass m times the acceleration l/t^2:

$$P = mlt^{-2}l^{-2} = ml^{-1}t^{-2}$$

These are the *physical dimensions* of pressure. The *unit of measure* of pressure in the CGS system is dyne/cm². However the units more frequently used in the medical field are the millimeter of mercury (mm Hg) and the centimeters of water (cm H_2O). In Fig. 1.2 are illustrated two manometers: a mercury manometer and a water manometer; the first one is used to measure high pressures, the second to measure small pressures. For example, to measure the pressure developed during a maximal expiratory effort it's used the mercury manometer, whereas to measure the smaller pressure developed during the elastic relaxation of the pulmonary system it's used the water manometer.

What is the relationship between the definition of pressure described above and the mm of mercury or the cm of water? In Fig. 1.3, it can be seen that if one applies a pressure on one branch of a manometer, the liquid inside (mercury or water) moves so that an unbalanced liquid column of height h is formed. This unbalanced column weighs (mg is the weight) on the surface of the cross section of the manometer:

$$P = mg/Section$$

Now let's introduce another physical dimension, the *density*. The density is the concentration of mass into the vacuum, i.e. $d = m/V$, different from the concentration, which is the amount of mass into the volume of another substance (for example glucose in water). Since the volume $V = Section\,h$, the pressure P can be defined as

$$P = mg/Section = dVg/Section = d\,Section\,hg/Section = dgh$$

This shows that the pressure P, measured by means of the manometer, *is independent of the section of the tube of the manometer*. Therefore, since $d = m/V = m/l^3 = ml^{-3}$ and the acceleration (velocity/time) $g = lt^{-2}$ the physical dimensions of the pressure are:

$$P = dgh = (m/V)\,gh = ml^{-3}lt^{-2}l = ml^{-1}t^{-2}$$

Since $P = d\,g\,h$ it is clear why it is expressed in mm of mercury or in cm of water thus defining the entity of the variables h and d. The only undefined variable is g, the acceleration of gravity on Earth. On the Moon where g is 1/6 that on Earth, the same pressure would result in a height h six time that on Earth.

The density of water is $d_{water} = 1$ g mass/cm³, that of mercury is $d_{mercury} = 13.6$ g mass/cm³.

Fig. 1.2 A mercury manometer (left) and a water manometer (right, water is colored). The right tube of each manometer is connected to the pressure source; the left tube is open to the atmospheric pressure

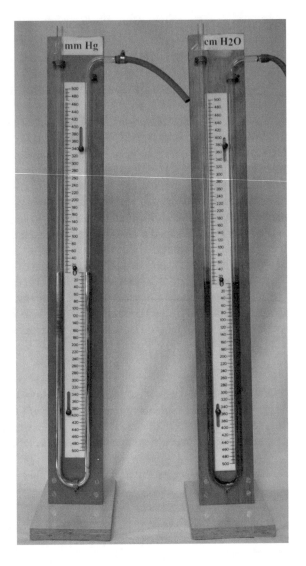

As mentioned above, blood moves from a point of greater pressure to a point of lower pressure (100–2 mm Hg in the systemic circulation). However is this statement always correct? The answer is NO. For example, in a glass of water standing on a table there is no motion of water in spite of the fact that the pressure of the water equals the atmospheric pressure Pb at the upper surface and $Pb + dgh$ at its bottom. Therefore this is one case in which in spite of a difference of pressure there is no motion of the fluid.

So the question is: what promotes the motion of blood? The drawing in Fig. 1.4 answers this question. Imagine an alpine lake above a valley. The pressure on the

Fig. 1.3 When a pressure P_{lung} is applied to one branch of the manometer (left in the Figure) an unbalanced column of fluid of height h takes place in the other branch (in communication with the atmospheric pressure P_b); the weight of this unbalanced column of fluid, applied perpendicularly to the section of the tube, counterbalances the applied pressure P_{lung}

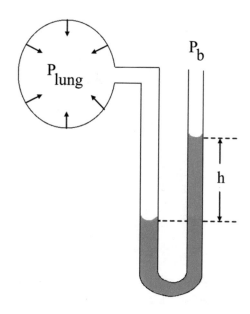

water surface is equal above and below (actually the atmospheric pressure is less above) and in spite of this water falls from above to below; this motion is due to a difference in gravitational potential energy, which is given by the weight mg times the height h, i.e. $E_p = mgh$. At the end of the fall, the water, which was still on the lake, moves forward with a velocity v due to the transformation of gravitational potential energy $E_p = mgh$ into kinetic energy $E_k = mv^2/2$. However, due to the friction on the soil, the velocity of water, i.e. its kinetic energy, progressively decreases so that, to continue its motion, it is necessary to pump the water forward, i.e. to impress a pressure P with a pump, to its volume V, adding the pressure energy PV (the energy impressed by heart to blood with its ventricular contraction).

In conclusion, what sustains the motion of a fluid is a difference in its mechanical energy, which is

$$E_m = mgh + mv^2/2 + PV$$

The mechanical energy per unit volume is:

$$E_m/V = dgh + dv^2/2 + P$$

I leave to you to show that the three terms of the above equation have the dimensions of a pressure: $ml^{-1}t^{-2}$. When blood flows from left ventricle to right atrium the difference in height h from the beginning to the end of the journey is about the same and the velocity is zero in both cavities; it follows that motion from start to end of the systemic (and pulmonary) circulation is sustained by a difference in pressure P only.

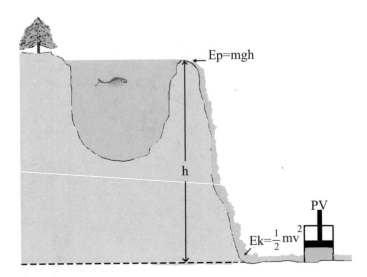

Fig. 1.4 This figure shows the three factors affecting the flow of blood: a difference in gravitational potential energy (E_p), in kinetic energy (E_k) and in pressure (P)

1.3   Statics of Circulation

1.3.1   Effect of Height

Statics is the study of pressure changes of blood taking place in the absence of its motion, i.e. with $v = 0$. Since motion is sustained by a difference in mechanical energy E_m, in the statics of circulation, where motion is nil, E_m will be the same in all vessels, i.e.

$$E_m/V = d\,g\,h + P = \text{constant}$$

and, consequently,

$$\Delta(E_m/V) = d\,\Delta g\,h + \Delta P = \Delta \ \text{constant} = 0$$

which represents the effect of acceleration on blood pressure P at a given quote, h; and:

$$\Delta(E_m/V) = d\,g\,\Delta h + \Delta P = \Delta \ \text{constant} = 0$$

which represents the effect of quote on blood pressure P at a given acceleration, i.e.

$$d\,g\,\Delta h = -\Delta P$$

Δh is positive when it is counted from below to above and negative from above to below. For example from heart to head dh is positive and, as a consequence ΔP is negative. On the contrary, from heart to feet dh is negative and as a consequence ΔP is positive. As mentioned above, in Statics the mechanical energy is equal in all sites of the circulation, in the head as in the feet. At all levels the sum $dgh + P$ is constant; for this reason water will not move in a glass standing on a table; similarly blood does not 'fall' from heart to feet, nor must be 'lifted' from heart to head as one would think. In Statics, blood has the same capability to move at all levels of its circle. Of these two forms of mechanical energy, however, P has the nasty physiological effect to dilate the vessels. If the vessels were rigid no physiological difference would be between a subject lying down and a subject standing upright.

Now let's calculate how the arterial pressure (suppose 100 mm Hg) would change because of the difference in height between heart and head (suppose 50 cm) and between heart and feet (suppose 130 cm). Since the difference in quote is measured in centimeters, the corresponding changes in pressure will be expressed in cm of water, cm H_2O (assuming the density of blood $= 1.055$ g cm^{-3}, equal that of water $= 1.0$ g cm^{-3}). However usually the pressure is measured in mm Hg. The density of mercury is 13.6 g cm^{-3}. How can we translate cm H_2O into mm Hg? This is explained below using the system CGS: Centimeter, Gram-mass and Second. The pressure dgh of 1 mm Hg will be:

$$dgh = 13.6 \times 980 \times 0.1 \approx 1300 \, \text{dyne cm}^{-2} \approx 1.3 \, \text{g} - \text{force cm}^{-2} \approx 1.3 \, \text{cm of blood}$$

where the density of mercury is $d = 13.6$ g-mass cm^{-3}, $g = 9.8$ m s$^{-2} = 980$ cm s^{-2} and $h = 1$ mm, i.e. 0.1 cm. Remember that dgh has the dimensions of pressure $ml^{-1}t^{-2}$. In conclusion the pressure of 1.3 cm of blood would equal that of 1 mm of mercury.

From heart and head we have assumed 50 cm, the height increases, so the pressure decreases:

$$1 \text{ mm Hg} : 1.3 \text{ cm blood} = x \text{ mm Hg} : 50 \text{ cm blood}$$
$$x = 50/1.3 = 38 \text{ mm Hg}$$

A difference of height of 50 cm of blood corresponds to a difference of 38 mm Hg. Therefore, assuming an arterial pressure of 100 mm Hg at the level of the heart the arterial pressure due to the difference in quote only will be $100 - 38 = 62$ mm Hg at the level of the head. At the level of the feet (130 cm below the heart in our example), the increase in arterial pressure will be $130/1.3 = 100$ mm Hg, i.e. the arterial blood pressure due to the difference in quote only will be $100 + 100 = 200$ mm Hg.

Let's consider now the venous pressure. Assuming a venous pressure of 2 mm Hg at the level of the heart we will have a pressure of $100 + 2 = 102$ mm Hg at the level of the feet. At the level of the head the venous pressure will be $2 - 38 = -36$ mm Hg: the venous blood pressure relative to the air outside will be negative! If you cut a vein of the neck in upright position air may be sucked within the vessel instead of blood spilling outside with serious consequences (embolism). However, if the veins of the neck collapse, how can blood flow into them? To understand this, imagine pressing a rubber tube: two channels will remain open in both sides allowing the flow into them. Inside the skull the venous vessels cannot collapse because they are contained in a rigid structure and are surrounded by incompressible structures (tissues) whose volume cannot change. Since the vessels of the lower limbs are extensible, the higher pressure due to their difference in quote when the subject is in an orthostatic position (200 mm Hg in the arteries and 102 mm Hg in the veins) will dilate the vessels of the legs particularly if the subject is stands still and the vessels are not squished by muscular contraction. As a consequence less blood will return to the heart, the blood emitted at each systole by the left ventricle will decrease, less oxygen will be brought to the brain and the subject may faint and fall to ground (*orthostatic hypotension*). When this happens, the initial cause of the problem will disappear because the body will lie horizontally on the ground. It is therefore advisable, in this case, to let him down instead of trying to lift him up. Fainting in a phone box may lead to death (Fig. 1.5).

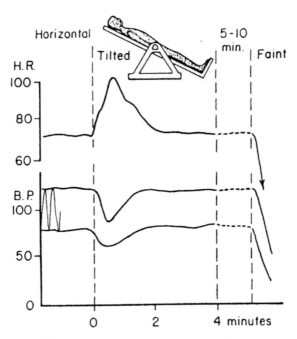

Fig. 1.5 Orthostatic hypotension evidenced in laboratory conditions. A subject is lying over a table (*tilting table*) in a horizontal position with his muscles relaxed: in this condition his heart rate (H.R. ~ 70 heartbeats/min) and blood pressure (B.P. ~80–120 mm Hg) are normal. At the time 0 (first interrupted line) the table is tilted and the subject is set in a vertical position. It can be seen that: (*i*) pressure falls because blood stagnates in the lower limbs, (*ii*) the heart initially reacts increasing the frequency of its contractions, but after 5–10 min heart rate and blood pressure fall and the subject faints (from Burton AC Year Book Medical Publisher, Inc., 1968)

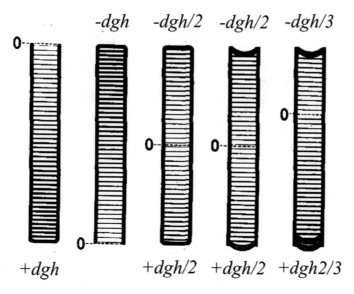

Fig. 1.6 Hydrostatic pressure at the ends of a fluid filled tube in the vertical position (modified from Clark JH, Hooker DR and Weed LH, *Am. J. Physiol.* 109: 166, 1934)

When muscles of the lower limbs contract (as in walking and running) the vessels are compressed and this promotes blood flow towards the heart because valves in the veins prevent motion in the opposite direction: this is called *muscular pump*.

Now let's consider another related argument: the *hydrostatic indifferent point* (HIP). Till now we have measured the static pressure changes due to a change in height above and below the heart. However, is it correct to assume the height of the heart as a reference level? To understand this point let's consider Fig. 1.6.

In the first picture on the left of Fig. 1.6 we have a tube of height h, full of water open at its upper extremity. It is clear that the pressure relative to the atmosphere will be zero at the top (as indicated) and $+dgh$ at the bottom of the tube due to the weight of the column of water (see Sect. 1.2). The opposite is true in the second picture where the tube is tilted upside down (assuming no water spills out of it): in this case the relative pressure will be zero at the bottom where the water is in contact with the atmosphere and negative, i.e. $-dgh$ at its upper extremity due to the weight of the column of water pulling down. Now let's close both ends of the tube; the level where relative pressure is zero (HIP) is set half way between the two extremities: pressure will be $+dgh/2$ at the bottom and $-dgh/2$ at the top. Why? Suppose to close the two extremities of the tube with two extensible rubber diaphragms instead of inextensible rigid walls. Even in this case $dgh = 0$ half way between the two extremities (fourth picture in Fig. 1.6). However if we close the upper extremity with one extensible diaphragm and the lower extremity with two extensible diaphragms, obtaining a stiffness two times greater at the bottom than at the top (right picture of Fig. 1.6), $dgh = 0$ moves upwards at a level where the height of the water column above it is 1/3 of the total height h, i.e. $-dgh/3$, and that below it is 2/3 of h, i.e. $+dgh2/3$. This finding can be understood as follows: since the total volume of liquid is constant, the volume of liquid moved inward at the

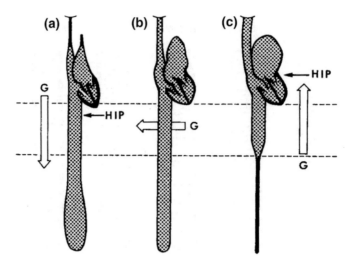

Fig. 1.7 Picture on the left (**a**) shows the upright position as indicated by the gravity vector G directed downwards (1 G of positive acceleration). Middle picture (**b**) shows the condition of a subject lying in a horizontal position. Right picture (**c**) shows the subject in up-side-down position (from Gauer OH and Thron HL *Postural changes in the circulation* in Hamilton VF and Dow P (eds) Handbook of Physiology, Vol. 3, Sect. 2, American Physiological Society, Washington, D.C., 1965)

top of the tube, due to the inflexion of the single layer membrane, *equals* the volume of liquid accepted at the bottom of the tube thank to the extension of the double layer membrane moving outward at the bottom of the tube. The deformation of the top and bottom membranes is the same, but to sustain this deformation, a pressure, i.e. the weight of the column of liquid counteracting the double layer membrane pushing upward at the bottom of the tube, due to its double stiffness, must be twice the column of liquid pulling downward on the single layer membrane. If we set the tube upside down, the position of $dgh = 0$ will remain at the same distance from the rigid and the extensible membrane even if the pressures at the tube extremities are opposite (i.e. it will maintain the same *relative position within* the tube); for this reason its position is called hydrostatic indifferent point, HIP. Figure 1.7 shows where HIP is placed in our body in the upright position (a) and in the head-down posture (c).

It can be seen that in the upright position (a) the hydrostatic indifferent point HIP is placed just below the heart ventricles. The middle picture (b) shows that in the horizontal position (1 G of transversal acceleration), when pressure is equal all-over the body (in static conditions), there is no appreciable difference in quote of the different body structures and therefore the HIP concept does not apply. Note that in this condition the heart is more extended than in the upright position: the volume of ventricles is maximal in the recumbent position (c). This is important because, as we will see (Sect. 1.6.1), the active tension that the muscular cell must exert for a given pressure increases with the radius of the container: the heart of a cardiopatic patient therefore will be in a better condition when sitting on the bed than when lying down. In the head-down posture (1 G of negative acceleration) HIP is placed just above

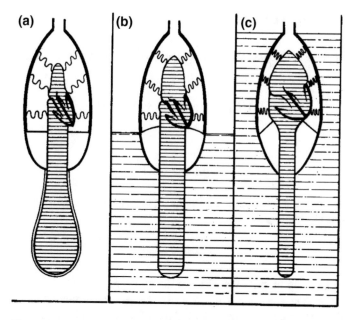

Fig. 1.8 a Shows the upright condition in air; **b** shows immersion in water to the diaphragm level and **c** to the neck (from Gauer OH, *Deut. Med. J.* 6:462,1955)

the ventricles and the blood of the legs is accommodated in the intra-abdominal region and in the intrathoracic region. In conclusion: the meaning of the hydrostatic indifferent point HIP is that its position within the body (i.e. relative to the body structures) does not change, or changes little (compare a and c in Fig. 1.7) when the body is tilted upside-down.

Now let's see what happens when we walk in the sea to take a bath (Fig. 1.8).

In the upright position in air (a) blood distends the veins of the lower limbs and the heart is relatively small because its position is above the HIP (Fig. 1.7). If we walk into water until its level attains the diaphragm (b), there is no longer swelling of the lower limbs because the increasing water depth implies an increase in pressure equal inside and outside the blood vessels (since, as we have seen, the density of blood practically equals that of water). It follows that the blood swelling the vessels of the lower limbs in air (a) will be pushed in the vessels above the water level with the consequence that the volume of the heart will increase and the greater pressure on the abdomen will push up the diaphragm. If we walk further into the sea until water level attains the neck (c) the pressure of water increases further whereas the average pressure of the air in the lungs remains on average equal to the atmospheric pressure, Pb, because the inside of the lungs is open to the outside. As a consequence the diaphragm will be pushed upward and even more blood will enter in the thoracic vessels causing a further increase of the volume of the hearth, which is surrounded by Pb.

1.3.2 *Effect of Acceleration*

Till now we have considered the effect of the quote assuming the acceleration g constant:

$$\Delta(E_m/V) = dg\,\Delta h + \Delta P = 0$$

i.e.

$$dg\,\Delta h = -\Delta P$$

Let's now consider the effect of a change in the acceleration g at a given quote on the statics of circulation, i.e.:

$$d\,\Delta gh = -\Delta P$$

The accelerations amplify the effect of the quote, h.

The acceleration is called *positive* when the subject starts moving upward and the blood tends to accumulate in the lower regions of the body (Fig. 1.7a): when upright we are subjected to 1 g of positive acceleration due to gravity. The acceleration is *negative* when the subject starts moving downwards and the blood tends to accumulate in the upper regions of the body (Fig. 1.7c) and *transverse* when acceleration acts perpendicularly to the major axis of the body (Fig. 1.7b). Tolerance is maximal for the transverse acceleration, intermediate for the positive acceleration and minimal for the negative acceleration. Relevant positive accelerations take place, for example, when a missile accelerate upwards at the start of its journey or when the pilots revert the motion of the plane from downward to upward; since the pressure in the eye is about 30 mm Hg, the transmural pressure of the retinal vessels is lower than the transmural pressure of the vessels supplying blood to the brain with the consequence that black-out may precede loss of consciousness in case of high positive acceleration. The effect of acceleration depends on its duration. Accelerations lasting less than 1 s (as an impact on the ground after a fall) have little effect on the circulation, i.e. on the distribution of blood volume within the vessels; their effect is to deform the body, i.e. to change the relative position of adjacent structures (e.g. air vs. alveolar tissue). In fact, according to the Archimedes principle, the net force F acting on a mass immersed in a fluid depends on the different density of adjacent structures:

$$F = mass\ body\ g - mass\ fluid\ g = volume\ g(density\ body - density\ fluid)$$

It follows that F, and as a consequence the relative displacement and the lesion of an organ due to an abrupt increase of g, depends on the difference in density between adjacent tissues and environment (maximal between lung tissues and air).

1.4 Dynamics of Circulation

1.4.1 Laminar Motion: The Law of Poiseuille

Over the pressure changes described above in the Static of Circulation, other pressure changes overlap due to the motion of blood. The Dynamics describes the pressure changes taking place in the circulation due to the motion of the blood caused by the contraction of the heart ventricles. To understand this motion we must understand what is a *fluid*. A fluid is substance that does not resist even to an infinitesimal small force tending to deform it, i.e. to slide one layer of its molecules over the adjacent one. However, even if infinitesimal, a force is required to slide a layer against the other; this is because the Brownian motion of the molecules interferes with the relative motion of layers. Figure 1.9 shows the force F necessary to slide two layers of surface A set at a distance Z. This force will be proportional to the surface A because the number of molecules interfering with the relative motion of the two layers will be greater the greater their number, i.e. the area A. Due to the limited range of the motion of the molecules interfering with sliding of the two adjacent surfaces, the force F will be smaller the greater their distance Z. Finally the force F will be greater the greater the velocity v of sliding of the layer relative to the adjacent one because the impact between molecules will increase with velocity. It is therefore understandable the equation

$$F = \eta A v / z$$

where the proportionality constant η is the *viscosity coefficient*, which is characteristic for each substance. The physical dimensions of η are $m\,l^{-1}t^{-1}$ as you can easily derive from the dimensions of each term in the above equation. The unit of measure in the

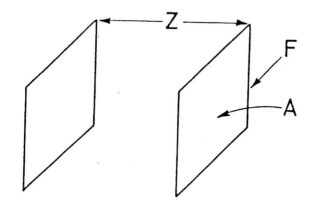

Fig. 1.9 Two surfaces of a fluid of area A, distance Z, sliding one over the other pushed by the force F (from Bull HB *An Introduction to Physical Biochemistry*, F.A. Davis Company, Philadelphia, PA, 1964)

CGS system is the *poise*. Since the viscosity of water at 20.2 °C is 0.01 poise, it is commonly used the *centipoise*; in addition since viscosity decreases with increasing temperature similarly in blood and water, the unit commonly used is the *relative viscosity = viscosity of blood/viscosity of water.* Viscosity of blood is about four times greater than that of water depending on the concentration of red cells. The relative volume of red cells is measured as the hematocrit H:

$$H = (Volume\ red\ cells/Volume\ of\ blood)100 \approx 45\%$$

Note that the *viscosity* must not be confused with the *density*: for example mercury is very dense, but little viscous; honey is very viscous but less dense.

If we approach the two surfaces A in Fig. 1.9 to an infinitesimal distance Z, the equation above can be written as

$$F = \eta A dv/dz$$

The ratio dv/dz is called *gradient of velocity.*

Figure 1.10 shows that if we push a book on a table the pages slide one over the other. Assuming an infinitesimal distance between pages dz the increment of velocity of one page relative to the other is dv. The bottom page of the book does not move. The work done by the hand pushing the book is required to deform it, i.e. to slide a page against the other, not to slide the book on the table. Similarly the work done by the heart is required to slide blood layers relative to each other. The difference in pressure, i.e. the work done by the heart per unit volume of blood, is dissipated as heat due to friction between blood layers not between blood and wall of the vessels.

The layer at the center of the vessel has the maximal velocity because it is the sum of the velocities of sliding of the different layers relative to the wall of the vessel where the velocity of the layer attached to it is zero. This ordered motion of the fluid is called *laminar motion.* The velocity decreases from the center of vessel to its wall as described in Fig. 1.11 with a parabolic trend that is called *parabolic profile of the velocities.* This trend can be described mathematically as follows. Let's imagine an infinite number of cylinders of base πz^2 and length l sliding one over the other like the cylindrical sections of a hand telescope. The force sustaining their motion will be $F = \Delta P \pi z^2$ where ΔP is the difference in pressure sustaining the motion of the blood within the vessel. The area of the cylinders where friction takes place is $A = 2\pi z l$. The equation we have derived above ($F = \eta A dv/dz$, see Fig. 1.9) can therefore be written as:

$$\Delta P \pi z^2 = \eta 2\pi z l(-dv/dz)$$

or

Fig. 1.10 Above: the velocity v of the layers of a fluid separated by the infinitesimal distance dz, increases relative to each other by the amount dv (modified from Edser E *General Physics*, London, Macmillan, 1926). Below: this is compared to a push by a hand causing deformation of a book, i.e. a sliding of its pages relative to each other, resulting in a velocity maximal at its upper surface and nil at its bottom surface due to adhesion with the table

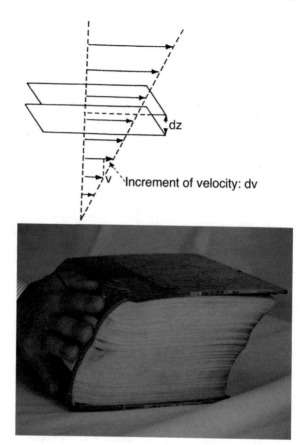

Increment of velocity: dv

$$dv/dz = -(\Delta P/2\eta l)z$$

Note that since dv/dz is negative because we have chosen as origin of axis the center of the vessel (Fig. 1.11), $-dv/dz$ and the force $F = \Delta P\pi z^2$ sustaining the motion are positive numbers.

The equation above, showing that the gradient of velocity $-dv/dz$ increases linearly with the distance z from center to wall of the vessel, can be rewritten as:

$$dv = -(\Delta P/2\eta l)zdz$$

and by integration ($\int$) of the linear relationship between $-dv/dz$ and z (imagine the indefinite area of a *triangle* given by the sum of the areas of an infinite number of *rectangles* having base dz and height $-dv/dz$)

$$v = -(\Delta P/2\eta l)(z^2/2) + \text{const} = -(\Delta P/4\eta l)z^2 + \text{const}$$

Fig. 1.11 The parabolic
profile of the velocity of the
layers of fluid in a tube (see
profile of pages at the bottom
of Fig. 1.10). The increment
of velocity per unit distance
(*dv/dz*) is maximal near the
wall of the tube

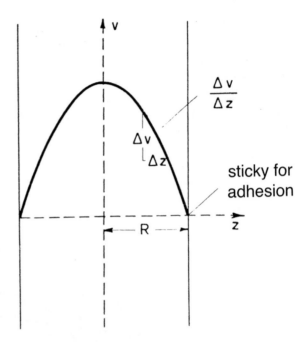

As described above the velocity $v = 0$ in contact with the wall of the vessel
where the distance z from the center of the vessel is R (i.e. the radius of the vessel,
Fig. 1.11). It follows that we can define

$$\text{const} = (\Delta P/4\eta l)R^2$$

i.e.

$$v = (\Delta P/4\eta l)\,(R^2 - z^2)$$

which represents the *parabolic profile* of blood velocity during *laminar motion* within
the vessel (Fig. 1.11).

The maximal velocity is at center of the vessel ($z = 0$).

$$v_{max} = (\Delta P/4\eta l)R^2$$

whereas the maximal gradient of velocity is adjacent the wall of the vessel where the
velocity is zero.

Instead of the *velocity* of blood in circulation and air in respiration, it is more
often used the *Flow*:

$$Flow = Volume/time$$

Since the *Volume* of fluid considered is that of cylinders of length l and base $A = \pi z^2$ the flow in each infinitesimal cylinder will be

$$\mathrm{d}\,Flow = v\mathrm{d}A = (\Delta P/4\eta l)(R^2 - z^2)\,2\pi z\mathrm{d}z$$

from which, by integration and rearranging, one obtains the *Law of Poiseuille*:

$$\Delta P = (8l\eta/\pi R^4)Flow$$

The term $(8l\eta/\pi R^4) = \Delta P/Flow$ is the *Resistance* offered to the flow of the fluid into the vessel. The term l/R^4 is called the *geometric* factor of the resistance.

In conclusion:

(1) The flow of blood in the systemic circulation equals that in the pulmonary circulation, i.e. the volume of blood emitted by the left ventricle in unit time equals that emitted by the right ventricle. However the *Resistance* of the systemic circulation is about 6 times greater than that in the pulmonary circulation: the average pressure in the aorta is about 100 mm Hg whereas that in the pulmonary vein is about 15 mm Hg.

(2) The fact that according to Poiseuille law, ΔP is related to the *fourth* power of the radius shows an amplification effect, i.e. small variation of the radius of the vessels result in large variation of the pressure. As we will see, at the levels of the *arterioles*, which represent the greater resistance of the systemic circulation, small changes of the radius cause large changes of ΔP, i.e. a fine regulation of the pressure is allowed.

(3) The motion, and as a consequence the loss of energy by friction, takes place between layers of blood not against the walls of the vessel; it follows that the level of smoothness of the walls is irrelevant, whereas what is relevant is the *viscosity* of the blood, which, as we will see, is greater the greater the concentration of the blood cells (i.e. the greater is the *hematocrit*). Subjects living at high altitude on the mountains have a greater concentration of blood cells and as a consequence a greater capability to transport oxygen, but their heart must exert a greater work against friction. The contrary is true for an anemic subject with a lower concentration of blood cells.

(4) Figure 1.11 shows that the maximum gradient of velocity $\mathrm{d}v/\mathrm{d}z$, and as a consequence the maximum work to overcome friction between layers of blood, takes place near the walls of the vessel (not *against* the walls).

Whereas the *plasma*, having a viscosity of 1.6 centipoise (water is 1 centipoise), can be considered a Newtonian fluid obeying the physical laws described above, the human blood, having a viscosity of 3–4 centipoise due to the presence of red cells deviates from the behavior of a Newtonian fluid in two ways: the *axial accumulation of red cells* and the *Fahraeus-Lindqvist effect*.

1.4.2   Axial Accumulation of Red Cells and Its Consequences

Imagine a volume of blood with a uniform distribution of red cells, i.e. with the same distance between each cell; now imagine setting two walls within this volume. It is clear that some red cells will interfere with the walls and could not stay there, and other will hit the walls and bounce away towards the center of the vessel. This is one reason for the axial accumulation of red cells. Another reason is explained in Fig. 1.12.

Imagine dipping a cylindrical stick in the center of a bucket full of water. Now imagine rotating the stick on itself. You will see that the water within the bucket will begin to rotate in the same way. This is because the layer of water adhering to the surface of the stick will transmit its motion to the adjacent layers. Red cells rotate into the vessel because plasma pushes on them with a velocity increasing towards the center of the vessel (Fig. 1.12). By rotating they increase the velocity of the plasma v_A adhering to them at center of the vessel and decrease its velocity near the wall v_B, i.e. $v_A > v_B$. According to the Bernoulli effect (stating that an increase in velocity of a fluid results in a decrease in its pressure), the pressure on the red cell near the wall will be greater than that near the center and this difference in pressure $P_r = P_r(B) - P_r(A)$ will contribute to the displacement of the red cells towards the center of the vessel. This process continues until the axial accumulation of the red cells opposes, as a wall, a further concentration of red cells in the center of the vessel. As we have seen (Fig. 1.11) the greater velocity gradient is near the walls of the vessel; the axial accumulation of red cells, by reducing the concentration of red cells in the plasma near the walls, decreases the viscosity of the moving fluid *just where the gradient of velocity* dv/dz *is greater*: this decreases the work done by the heart to sustain the motion of blood, which, as described above, is spent to overcome friction between blood layers (remember that $F = \eta A dv/dz$). The *axial accumulation of red cells* described above implies that the red cell concentration within the plasma of a blood vessel sprouting laterally from a larger artery may be lower than required.

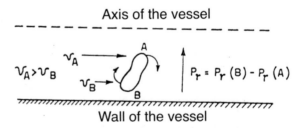

Fig. 1.12 As shown in Fig. 1.11 the velocity near the center of the vessel v_A is greater than that near the wall v_B. This results in a rotation of the blood cell and of the plasma adhering to it that, as described below, causes a displacement of the red cell near the center of the vessel (modified from Burton AC, *Hemodynamics and the physics of the circulation*, in Ruch TC and Patton HD (eds), *Physiology and Biophysics*, W.B. Saunders Company, 1966)

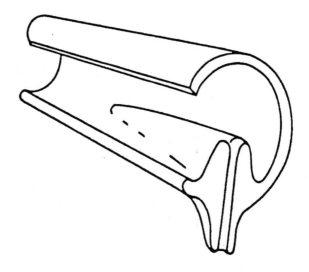

Fig. 1.13 Schematic representation of an arterial cushion withdrawing blood from the center of the vessel where the red cells concentration is higher (modified from Fourman J and Moffat DB, *J. Physiol.* 158: 374, 1961)

This problem is solved as indicated in Fig. 1.13 by *arterial cushions* drawing blood from the center of the artery where the concentration of blood cells is higher.

1.4.3 The Fahraeus-Lindqvist Effect and Its Consequences

Figure 1.14 shows the relative viscosity of blood (*viscosity of blood/viscosity of water*) as a function of the *hematocrit (Volume red cells/Volume of blood)* measured in two different conditions: within a *viscosimeter*, i.e. a tube having a diameter of 1 mm, and in the *arterioles* having a diameter of ~0.1 mm. It can be seen that the viscosity of blood increases with the hematocrit differently in the two conditions. At the normal hematocrit of about 45% (i.e. 45 ml of red cell volume in 100 ml of blood) the relative viscosity is about 4 when measured with the viscosimeter and about 2 when measured in the arterioles (*Fahraeus-Lindqvist effect*).

Figure 1.15 shows that the same effect is observed with glass tubes of different radius: the relative viscosity of blood decreases with the radius of the tube from its true value remaining constant when measured with tubes of increasing size.

There are two explanations for this finding. The first refers to the fact that the ratio *surface/volume* of the vessel is greater in the small vessel than in the larger one; this is because by decreasing the size of the vessel the volume decreases with the *cube* of its linear dimensions, whereas the surface of its cross-section decreases with the *square* of them. It follows that the surface nearest to the vessel's walls, which as described above has a lower concentration of red cells, and as a consequence a lower viscosity, is relatively greater in the vessels of smaller dimensions.

The second explanation is the *sigma effect* (Σ) referring to the fact that when the dimensions of the red cells become appreciable relative to the cross section of the

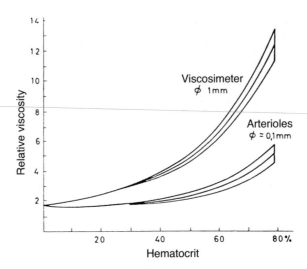

Fig. 1.14 Relative viscosity of blood (ordinate) as a function of the red cells concentration (Hematocrit, abscissa). The upper curve was obtained with a viscosimeter having the diameter of 1 mm, the lower curve was obtained on the blood vessels of the leg of a dog (the average diameter of the arterioles, offering the greatest resistance, is less than 0.1 mm). On each side of the curves is indicated the standard deviation. The influence of the radius of the vessel on the apparent value of viscosity is discussed in the text (modified from Ruch TC and Patton HD (eds), *Physiology and Biophysics, W.B. Saunders Company*, 1966, from the data of Whittaker SRF and Winton FR, *J. Physiol.* 78: 339, 1933)

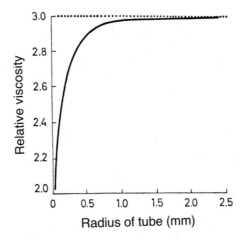

Fig. 1.15 *Fahraeus-Lindqvist effect*. The dimensions of the glass tube in which blood flows affect drastically the value of viscosity calculated according the law of Poiseuille (ordinate) when the radius (abscissa) decreases below 1 mm. The dotted horizontal line indicated the real value of viscosity, coinciding with that calculated for all tubes of radius greater than ~2 mm (modified from Burton AC, *Physiology and Biophysics of the Circulation*, Year Book Medical Publisher, 1968)

vessel, the layers of plasma behind them will not slide relative to each other. As a consequence instead of friction between an infinite number of layers, assumed when deriving by integration ($\int$) the law of Poiseuille as described above, the total friction to overcome will be only the sum (Σ) of a finite number of layers moving behind red cells without friction within them. If the diameter of a red cell would equal that of the vessel no work would be necessary to overcome frictions within the plasma behind it because the velocity of each layer of plasma behind the red cell would be the same.

1.4.4 Turbulent Motion

Contrary to the laminar motion where the layers of blood slide parallel each other and to the wall of the vessel, in the turbulent motion some layers of blood twist on themselves, moving perpendicularly against the walls and backwards against the main flow. This leads to the formation of *vortices*, i.e. large differences in direction and blood velocity within the vessel. Their pulsations against the wall of the vessel may create noise, which can be detected by auscultation with a phonendoscope. The laminar flow is silent, the turbulent flow is noisy: you may stay near a river without hearing its flow, whereas you may hear the noise of a creek at a distance from it. The vortices imply continuous acceleration and deceleration of the fluid, i.e. changes in its kinetic energy and work to sustain them. This additional work adds to the work done against friction making the turbulent flow more expensive than the laminar flow. As described above in the laminar flow the driving pressure increases linearly with flow according to the law of Poiseuille, where the *resistance* $= \Delta P/Flow = (8 \, l\eta/\pi R^4)$ is constant, i.e. the relationship between P and *Flow* is a straight line. In the turbulent flow, on the contrary, the relationship between driving pressure and flow, $\dot{V}$, is not linear: $\Delta P = a\dot{V}^x$ where x is greater than 1 and less or equal 2; in the turbulent flow the slope $\Delta P/Flow$ increases with flow more or less steeply according to the value of x and the *scatter of the data is greater than in the laminar flow.*

In a tube of a given geometry, the laminar flow becomes turbulent when the velocity of the fluid increases above a value called *critical velocity.* Since the volume V of a cylinder is given by the product of its sectional area times is length l, i.e.

$$V = Section \; l$$
$$V/time = Flow = Section \, l/time = Section \times velocity$$
$$velocity = Flow/section$$

Now, let's derive logically the value of the *critical velocity*, v_{crit}. Let's consider the physical properties of the fluid and the geometry of the vessel in which the fluid flows. As mentioned above the turbulent flow creates *vortices*: the layers of fluid tend to separate from one another resulting in a zone of negative pressure between them called *cavitation*. This phenomenon is evident when by shaking a sparkling drink the gas in solution sets free in bubbles. The characteristics of the fluid to consider in this case are its viscosity η and its density d. As described above, viscosity η can be viewed as links between the layers of fluid increasing the frictional drag in the laminar motion and opposing divergence between layers, i.e. cavitation, as in the turbulent motion. Honey is more viscous than water and it is intuitive that bubbles will form more easily by shaking water than honey. We can then conclude that the critical velocity necessary to attain turbulence must be greater the greater the viscosity, i.e. that v_{crit} increases with η. *Vortices* imply deceleration and acceleration of particles of fluid pushing and bouncing off the wall of the vessel. Accelerations, i.e. changes in velocity, imply changes in kinetic energy, $\frac{1}{2} mv^2$. The density of the fluid is $d = m/V$: it follows that kinetic energy changes will be larger the denser the fluid. Imagine a curvilinear path (as that of air through the nose during inspiration) the centrifugal force on the air layers and on the pollution particles within them will be greater the greater their mass, with the consequence that both will be projected against the wall, some of them usefully retained (the pollution ones) others (the air ones) bouncing back creating vortices i.e. turbulent motion. This example shows that curvature, narrowing, roughness of the vessel will tend to create vortices more likely the greater the mass per unit volume of the fluid, i.e. the greater its density. We can then conclude that the critical velocity necessary to attain turbulence v_{crit} will be lower the greater the density, d.

Now let's consider the geometry of the vessel. Curvatures, narrows, in general deviation from a straight, smooth cylindrical vessel favor *ceteris paribus* turbulent motion. But how the dimensions of a smooth cylindrical vessel affect outbreak of turbulent motion? Figure 1.16 (top) shows a device (the flowmeter of Fleisch), which decreases the insurgence of turbulent flow; it consists of a tube filled with small cylindrical tubes parallel to each other. This is because vortices need space to develop and their insurgence is delayed, i.e. occurs at a greater critical velocity, when the layers of fluid flow in tunnels of small dimensions. Figure 1.16 (bottom) shows the section of a flowmeter, more commonly used in practice, consisting in the insertion within the conduct of a very thin metallic mesh (interrupted line in Fig. 1.16, bottom), which helps an ordered laminar flow of the air. Flowmeters are very useful in practice because they allow determining flow from the drop of pressure ΔP when the flow is *laminar (Flow* $= (8 \, l\eta/\pi R^4)/\Delta P)$. In contrast flow is not determined reliably when the flow is turbulent due to irregular relationship between ΔP and flow in this condition.

From what described above we can conclude that the *critical velocity* v_{crit} above which turbulence appears will be greater the greater the viscosity of the fluid η, and smaller the greater the density of the fluid d and the radius of the vessel r, i.e.:

$$v_{crit} = N_c \eta/dr$$

The proportionality constant N_c is called the *critical Reynolds Number*, which is dimensionless. This is shown below: the dimensions of the variables v_{crit}, η, d and r in the above equation cancel out:

Fig. 1.16 In order to maintain an ordered laminar flow (*Law of Poiseuille*), the gradient of pressure ΔP is measured across a tube filled of small cylindrical tubes (top) or a thin metallic mesh (bottom interrupted line) both preventing turbulent flow

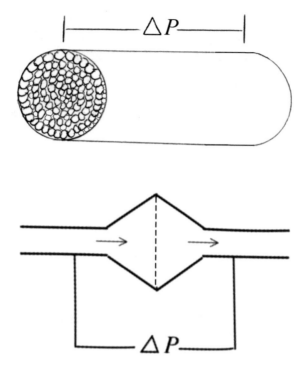

$$N_c = v_{crit} \times d \times r/\eta = (lt^{-1}) \times (ml^{-3}) \times (l)/ml^{-1}t^{-1})$$

When the flow takes place *in a smooth straight cylinder of constant sectional area* $N_c = 1000$ provided that the values of the variables are all expressed consistently with the same system of measurement (e.g. CGS: centimeter-gram-second). In the presence of curvatures, roughness or constrictions (Fig. 1.17) N_c decreases with the consequence that the critical velocity at which turbulence takes place is lower. This allows the doctor, by auscultation with a phonendoscope, to detect irregularities in the geometry of blood vessels and bronchi.

It is important not to confuse the radius of the overall section of capillaries (as indicated in Fig. 1.18 by the interrupted line) with the radius of each singular vessel. In the equation defining the critical velocity ($v_{crit} = N_c\eta/dr$), the radius r is that of the singular vessel where turbulence may occur: the smaller the radius the greater the critical velocity above which turbulence takes place. It follows that possibility of turbulence will decrease from the aorta to the capillaries (or from the trachea to the bronchioles) for two reasons: (*i*) because the average velocity of the blood (or the air) in each conduct decreases from center to periphery due to the increase of the overall cross sectional area where the flow is distributed (Fig. 1.18) and (*ii*) because the critical velocity v_{crit} in each conduct increases due to a reduction of r. For these reasons it is more likely to hear noise with a phonendoscope in proximity of the central large conducts than in the small peripheral conducts.

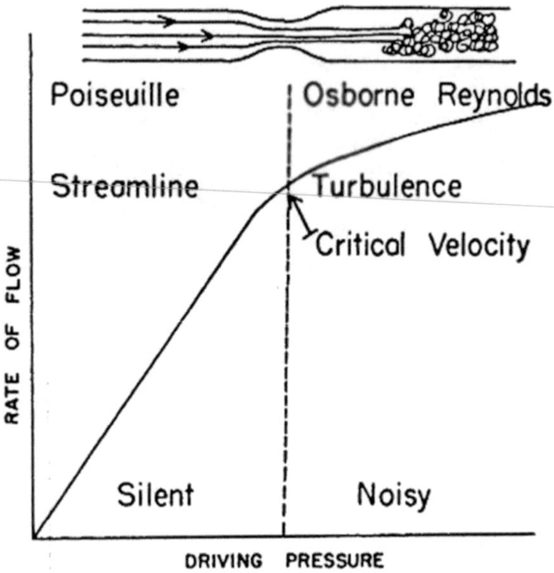

Fig. 1.17 Change in type of flow from laminar to turbulent at the critical velocity ($v_{crit} = N_c\eta/dr$). In the drawing above turbulence is caused by a local narrowing of the vessel (modified from Ruch TC and Patton HD, *Physiology and Biophysics*, Saunders Company, 1974)

Poiseuille ┊ Osborne Reynolds

Streamline ┊ Turbulence

┊ Critical Velocity

RATE OF FLOW

Silent ┊ Noisy

DRIVING   PRESSURE

Fig. 1.18 To show that the same blood flow in the aorta distributed in the larger comprehensive area of all capillaries results in a proportional decrease in blood velocity

velocity = flow/section

flow aorta   =   flow all capillaries

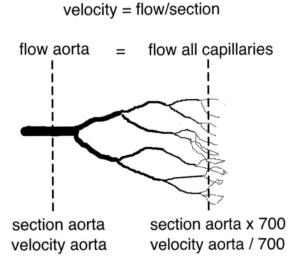

section aorta   section aorta x 700
velocity aorta   velocity aorta / 700

Now let's compare the critical velocity v_{crit}, above which the turbulence takes place, with the actual average velocity of blood in the aorta. Using the system CGS and assuming blood density $d = 1$ g mass/cm^3 (actually blood density is about 1.055 g mass/cm^3 depending on the hematocrit), the radius of the aorta $r \approx 1$ cm and blood viscosity $\eta = 0.04$ poise

$$v_{crit}(\text{cm/sec}) = 1000 \times 0.04 = 40 \text{ cm/sec}$$

in human at rest the cardiac output is ~5 l/min, corresponding to an average flow in the aorta of $5000/60 = 83$ cm^3/s and to an average velocity (flow/section) of $83/3.14 = 26$ cm/s. Thus, considering the *average* velocity, turbulence would not tale place not even at the level of the aorta. However during the systole the velocity of blood is greater than average with consequence that noise due to turbulence can be heard by a phonendoscope particularly when the cardiac output is increased as during exercise, in an excited patient or when the viscosity η is low due to a reduced number of blood cells in an anemic subject.

1.4.5   Incidence of Kinetic Energy in Circulation

As explained above (Fig. 1.4) the mechanical energy of a fluid is:

$$E_m = mgh + mv^2/2 + PV$$

The mechanical energy per unit volume is:

$$E_m/V = dgh + dv^2/2 + P$$

Let's assume a fluid moving in a horizontal tube so that $\Delta(dgh) = 0$ and neglect frictions so that $\Delta(E_m/V) = 0$, then

$$\Delta(dv^2/2) = -\Delta P$$

i.e. *a change in kinetic energy involves an opposite change in pressure*. If frictions are taken into account as in Fig. 1.19a, an increase in kinetic energy due to a narrowing of the tube, would lead to a fall in pressure K.E. greater than that caused by frictions. It can be seen from Fig. 1.19a that flow (velocity times area) is the same in each section of the tube; the total mechanical energy E, falls linearly downstream as expected, but the pressure is lower at the center of the tube than at the end of it, i.e. flow takes place *against* a positive gradient of pressure.

The incidence of kinetic energy must be taken into account when measuring blood pressure directly by means of a catheter as described in Fig. 1.19b. Blood pressure P is measured *correctly* provided that no interference would take place with its flow, i.e. laterally, perpendicularly to the flow, as indicated by the height of the column of fluid in the *piezometric tubes* in Fig. 1.19a, or by the lateral opening of the catheter in the middle drawing of Fig. 1.19b: *side pressure = P*. If on the contrary the opening of the catheter faces the flow stopping it, the loss of the kinetic energy term results in an increase in pressure (K.E.) beyond the correct side pressure P: *end pressure = P + K.E.* If the opening of the catheter faces opposite the direction of flow, whirls will take place sucking liquid out of the catheter with the consequence that the measured pressure, will be lower than the correct side pressure P: *downstream pressure = P −*

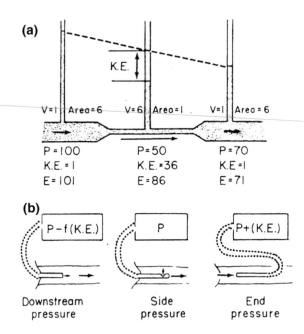

Fig. 1.19 a Since flow $=$ velocity $\times$ cross-sectional area, a narrowing of the tube results in a proportional increase in velocity v, in the kinetic energy K.E. $= mv^2/2$ and a decrease in pressure P. The total mechanical energy $E = P +$ K.E. decreases linearly all along the horizontal tube, but after the narrow, flow takes place against a positive gradient of pressure. **b** Illustrating possible artifacts when measuring blood pressure P with catheters: the correct measurement is made when the opening of the catheter is perpendicular to the flow (*side pressure*) avoiding interference with the velocity of the flowing blood (from Burton AC *Physiology and Biophysics of the circulation*, Year Book Medical Publishers Incorporated, Chicago, 1965)

f (K.E.) $\approx P - 0.8$ (K.E.). The error made by measuring incorrectly the *end pressure* or the *downstream pressure* instead of the correct *side pressure* will be greater where the blood velocity and, as a consequence, the kinetic energy term K.E., are greater, i.e. in the large vessels: aorta, pulmonary artery, cava veins and pulmonary veins. The *relative* error will be greater in the vessels where the *side pressure* is lower (e.g. in the pulmonary artery where the average $P \approx 15$ mm Hg and in the large veins where $P \approx 2$ mm Hg vs. ≈ 90 mm Hg in the aorta at rest). The error is greater the greater the kinetic energy of blood, i.e. the greater the cardiac output. As it will be described below, the velocity varies in the arteries from a maximum during the systole to a minimum during the diastole whereas in the veins the velocity is constant.

When the heart ventricles contract they must develop a pressure (*end pressure*) that must account for both the systolic pressure in the aorta and in the pulmonary artery *and* the kinetic energy in these vessels (*end pressure* $= P +$ K.E.). Let's consider the left ventricle and calculate how much the pressure in the left ventricle (*end pressure*) is higher than the pressure in the aorta (P) due to the pressure term K.E. Using the C.G.S. system:

$$K.E. = 1/2dv^2 = 1/2 \times 1 \times 30^2 = 450 \text{ dyne/cm}^2$$

where $d = 1$ g-mass/cm^3 (assuming the density of blood $= 1.055$ g-mass/cm^3, equal that of water $= 1.0$ g-mass/cm^3) and $v \approx 30$ cm/s is the average velocity of blood in the aorta at rest. However usually the pressure is measured in mm Hg. The density of mercury is 13.6 g cm^{-3}. As explained above (Sect. 1.3.1), the pressure $d\,g\,h$ of 1 mm Hg will be (using the system CGS):

$$dgh = 13.6 \times 980 \times 0.1 \approx 1300 \text{ dyne/cm}^2$$

where the density of mercury is $d = 13.6$ g-mass cm^{-3}, $g = 9.8$ m s$^{-2} = 980$ cm s^{-2} and $h = 1$ mm, i.e. 0.1 cm. It follows that

$$1 \text{ mm Hg} : 1300 \text{ dyne/cm}^2 = X \text{ mm Hg} : 450 \text{ dyne/cm}^2$$

$$X = 450/1300 = 0.35 \text{ mm Hg}$$

It follows that the additional pressure K.E. that the left ventricle must exert to increase the average blood velocity from zero (in the ventricle) to the average velocity in the aorta is a small fraction of the total average pressure measured in the aorta (≈ 100 mm Hg). However the calculation above results in a value of the K.E. pressure lower than real for two reasons: (*i*) because the square of the *average* velocity is smaller than the average of the squares of the instantaneous velocities, and (*ii*) because the velocity during the systole is higher than the average velocity. Taking into account these two factors one obtains that the K.E. pressure term attains ≈ 3 mm/Hg during cardiac output at rest. During maximum muscular exercise in an athlete the cardiac output may increase 5 times, i.e. the kinetic energy term may increase 25 times and the K.E. pressure term attains ≈ 75 mm/Hg. As a consequence the pressure in the left ventricle will attain during the systole a maximum value of $150 + 75 = 225$ mm Hg. In conclusion the pressure in the left ventricle during the systole with a maximal cardiac output will attain a value much higher than the pressure measured in the aorta. In the veins the error can range from $\approx 12\%$ at rest to $\approx 52\%$ when the cardiac output is increased three times (Table 1.1).

1.5 The Sphygmic Wave (Arterial Pulse)

The pressure and the velocity of blood in the aorta and in the pulmonary artery are not constant, attaining a maximum during the systole and a minimum during the diastole. This involves changes in kinetic energy and work to sustain them. Why blood flow is not maintained constant by a rotating device as, for example in the extracorporeal circulation? The answer is that rotating devices, such as a wheel, do not exist in macroscopically living organisms. In circulation, respiration and locomotion an expensive for-back contraction and relaxation of the heart, respiratory and limb

Table 1.1 Relative importance of kinetic energy in different parts of the circulation (*italic* figures show where kinetic energy is more than 5% of the total blood energy; Neg.= negligible)

Vessel	Resting cardiac output				Cardiac output increased 3 times		
	Velocity (cm/s)	Kinetic energy (mm Hg)	Pressure (mm Hg)	Kinetic energy as % of Total	Kinetic energy (mm Hg)	Pressure (mm Hg)	Kinetic energy as % of Total
Aorta, systolic	100	4	120	3%	36	180	*17%*
Mean	30	0.4	100	0.4%	3.8	140	2.6%
Arteries, systolic	30	0.35	110	0.3%	3.8	120	3%
Mean	10	0.04	95	Neg.		100	Neg.
Capillaries	0.1	0.000004	25	Neg.	Neg.	25	Neg.
Venae cavae and atria	30	0.35	2	*12%*	3.2	3	*52%*
Pulmonary artery, systolic	90	3	20	*13%*	27	25	*52%*
Mean	25	0.23	12	2%	2.1	14	*13%*

From Burton AC *Physiology and Biophysics of the circulation*, Year Book Medical Publishers Incorporated, Chicago, 1965

muscles takes place. This is because the rotating device must be connected to the organism by vessels to be nourished, by nerves to be controlled and both vessels and nerves would be twisted during rotation. Only in microscopically living organism connection can be assured by diffusion. Contrary to this intermittent flow of blood at the origins of arterial circulation, both systemic and pulmonary, the metabolism of the living cells (i.e. the O_2 and the CO_2 gaseous exchange and the input output of molecules and heat) is a continuous process necessary to maintain constant the internal environment. Accordingly the flow of blood in the capillaries, where the exchange takes place, is continuous in spite of its pulsatile discontinuity at its origin. How this happens? The transition of blood flow from discontinuous to continuous takes place thanks to two characteristics of the arterial tree: its *extensibility* or *compliance* and its *elasticity*. To understand qualitatively the difference between these two physical properties of a structure consider the difference between a steel ball and chewing gum: the first is rigid, little extensible, but very elastic; the second is very extensible but inelastic. Figure 1.20 shows how these two properties can be defined in physical terms.

The right hand line in Fig. 1.20a shows the pressure-volume relationship of a structure less stiff, i.e. more compliant, than that shown to the left. The area below the curves, i.e. the integral $\int PdV$ represents the elastic energy stored (if the relation-

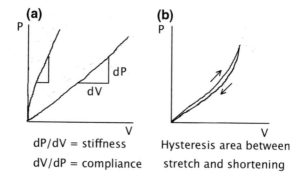

Fig. 1.20 The two graphs define the two characteristics of the arterial tree that allow transforming the intermittent flow in a continuous flow: its extensibility or compliance (**a**) and its elasticity (**b**)

ship is linear the area will be that of triangle, i.e. *PV/2*): from Fig. 1.20a it can be seen that the elastic energy stored at given volume is greater the greater the compliance of the structure (right hand line). The compliance is a characteristic of the arterial tree: its volume increases during the systole when the pressure increases and decreases during the diastole when the pressure decreases. The other characteristic is its elasticity, i.e. its capacity to restore most of the elastic energy stored during stretching. Figure 1.20b shows that the area below the P-V relationship is greater during stretching (upward arrow) than during shortening (downward arrow): the area between the two lines is the mechanical energy lost during the stretch-shortening cycle and is called *elastic hysteresis*; the smaller the hysteresis the more elastic is the structure. In conclusion: during the systole the increase in cross-sectional area of the arterial vessels tends to decrease the velocity of blood and stores elastic energy. During the diastole the decrease of cross-sectional area of the arterial vessels increases blood velocity due to the release of the elastic energy previously stored with the result that blood flow, instead of being intermittent as in a rigid inextensible tube, tends to become continuous in the capillary bed. In case of a local narrowing of a vessel, the increase in blood velocity, i.e. of its kinetic energy, may cause a fall in pressure resulting in a closure of the vessel and stopping of blood flow; as soon as the vessel close however, the end pressure upstream the narrowing will reopen it and so on. This condition is called a *flutter*. In case the upstream pressure is unable to reopen the vessel, e.g. a coronary artery, an infarct would occur. As mentioned above, blood pressure is not constant in the large arterial vessels due to the intermittent contraction and relaxation of the heart ventricles. In the *systemic* circulation these oscillations of pressure are easily felt at the level of the large arterial vessels, such as the radial artery (*radial pulse*). The oscillation in pressure in the large arteries of the systemic circulation is called the *sphygmic wave* (Fig. 1.21).

At each systole the left ventricle pumps in the arterial tree ~60–80 ml of blood (*cardiac output pulse*), which initially distends the arterial tree. This results in an increase in pressure (*anacrotic phase of the sphygmic wave*). However after attaining a maximum the pressure falls in the arterial tree in spite of a continuous injection of blood from the left ventricle into the aorta. This is because the flow input in the arterial tree from the ventricle is less than the flow output from the arterial tree into

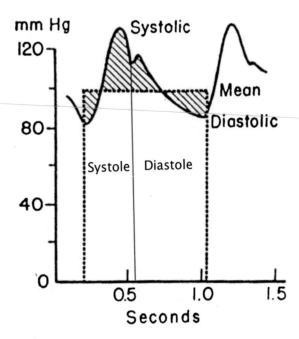

Fig. 1.21 Normal sphygmic wave at rest: the pressure oscillates from a minimum (Diastolic, ~80 mm Hg) to a maximum (Systolic, ~120 mm Hg). The mean arterial pressure (true mean or area mean) is calculated dividing the whole area below two minima of the sphygmic wave by the time interval between the two minima. The area of the curve above the height of the rectangle, so obtained, equals the sum of the two areas below. Note that the duration of the diastole is greater than that of the systole (modified from Burton AC *Physiology and Biophysics of the circulation*, Year Book Medical Publishers Incorporated, Chicago, 1965)

the capillaries through the *arterioles*: the peripherals resistance units (*PRU*) (crosses in Fig. 1.1). If the arterial tree were rigid, the pressure would fall to zero when the output pulse stops. On the contrary, blood is stored under pressure in the *extensible and elastic arterial tree* and tends to flow in both directions: forward through the PRU and backward in the left ventricle. The backward flow of blood into the left ventricle is prevented by the semilunar valves. Their loading and recoil results in an oscillation: the *dicrotic wave* (Fig. 1.21) which is less muffled when pressure is measured near the ventricle.

After the systole the pressure falls to a minimum value $P_{min} = P_{max} - \Delta P$ during the diastole. What are the factors determining the fall in blood pressure $-\Delta P$ from its maximum value P_{max} down to its minimum value P_{min}?

(1)   The time interval after which a motion resumes the same characters is called *period* (τ). The frequency of the motion is $f = 1/\tau$. In case of heart contraction and relaxation, $\tau = systole + diastole$ durations, with *diastole duration* > *systole duration* (Fig. 1.21). It is therefore clear that the fall in pressure $-\Delta P$ will be

greater the lower the frequency of heartbeats, i.e. the greater the time during
which the pressure falls after the systole.

(2) If the arterial tree were rigid the pressure would fall instantaneously to zero after
the systole. The more extensible and elastic is the arterial tree the lower will be
the velocity of the fall in pressure $-\Delta P/\Delta t$, and, for a given heart frequency,
the diastolic pressure will be higher.

(3) The blood stored in the arterial tree during the systole flows to the capillaries
through the arterioles. i.e. the peripheral resistance units (PRU, Fig. 1.1). It is
therefore clear that $-\Delta P/\Delta t$ will be lower and, for a given heart frequency, the
diastolic pressure will be higher the greater the peripheral resistance units, PRU.

Let's now consider what determines the maximal pressure $P_{max} = P_{min} + \Delta P$.

(1) $+\Delta P$ will be greater the greater the *cardiac output pulse*: e.g. if instead of 60 ml
of blood, the left ventricle injects 120 ml of blood in the arterial tree $+\Delta P$ will
be obviously greater.

(2) $+\Delta P$ will be lower the greater the *compliance* (not necessarily the *elasticity!*)
of the arterial tree. A rigid arterial tree will result in a greater $+\Delta P$.

(3) Less important in increasing $+\Delta P$ is an increased resistance of PRU, given the
short time available to drain blood during the systole.

The minimal and the maximal pressures attained by the blood in the systemic
arterial tree oscillate below and above the *mean arterial pressure*, which can be
defined in three ways:

(1) The pressure that maintained constant over the period τ of the sphygmic wave
has the same effect in promoting the flow of blood. Since the motion of blood
from ventricle to the atria is sustained, as described previously, by a difference
in mechanical energy per unit volume ($E_m/V = dgh + dv^2/2 + P$), and since
the difference in height h is about the same and the velocity v is zero at the
beginning and at the end of the journey, we can conclude that the *average
pressure $\bar{P}$ equals the mechanical energy per unit volume sustaining the blood
flow* from the left ventricle to the left atrium. It follows that the average pressure
must necessarily decrease along the path from ventricles to atria because it
represents a mechanical energy per unit volume that is progressively degraded
into heat by frictions along its path. This is in contrast with the systolic maximal
pressure that may even increase due to a deformation of the sphygmic wave (as
the height of the sea waves approaching the shore).

(2) Geometrically: the pressure value that maintained constant over the period τ
subtends the same area of the sphygmic wave; i.e. the sum of the two areas
below the mean pressure equals the area above it (dashed in Fig. 1.21).

(3) Mathematically. Even without a mathematical preparation, it is possible to
understand that the area below the sphygmic wave can be represented by the
sum ($\int$) of the areas of rectangles having an infinitesimal small increment of
time as abscissa dt and pressure p during this time; i.e. this sum is represented
by the *integral*: $\int P \, dt$. Since, as mentioned above, the area is given by the aver-

age pressure multiplied by the time period τ, it is possible to define the average pressure as $\overline{P} = \int P \, dt/\tau$.

The average pressure, and consequently the average flow, is greater in extensible and elastic vessels then in rigid vessels of the same cross section because in the rigid vessel more energy is spent in turbulent motion due to the higher velocity attained, whereas in the extensible and elastic vessel more energy is saved as elastic potential energy when their section increases during the systole and subsequently recovered when their section recoils during the diastole. In addition in a rigid tube the kinetic energy $\frac{1}{2} m v^2$ will fall from a maximum to zero at each systole with the consequence that more energy will be spent to sustain the kinetic energy changes of blood (as driving a car in the city instead of on the highway).

The sphygmic wave changes in different pathologies as indicated in Fig. 1.22.

It can be seen that in the aortic stenosis (a bottleneck at the level of aorta), the maximal systolic pressure is decreased and substituted by an interval of lower pressure and noisy turbulence. In the arteriosclerosis, the compliance of the arterial tree is decreased and, as described above, $+\Delta P$ is increased with the consequence that the pressure raises to a high value during the systole and subsequently falls with a high velocity as in a rigid tube; the higher minimal pressure is due to an associated increase of the PRU. In the aortic insufficiency blood during the diastole flows not only forwards through the PRU, but also backwards into the left ventricle: this explains the large fall in pressure to a minimum value and the subsequent greater $+\Delta P$ due to the additional amount of blood to be emitted during the systole. The sphygmic wave, that we can feel for example on the radial artery at the wrist, is characterized by its:

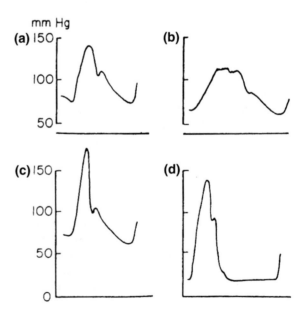

Fig. 1.22 a Normal. **b** Aortic stenosis. **c** Arteriosclerosis. **d** Aortic insufficiency (from Burton AC *Physiology and Biophysics of the circulation*, Year Book Medical Publishers Incorporated, Chicago, 1965)

(1) *Frequency* $= 1/\tau$. Note that when you measure the frequency of the pulse beats you must begin the count as follows: 0, 1, 2, 3 and so on, i.e. 3 intervals in this case, not as 1, 2, 3, 4, thus counting wrongly 4 intervals instead of 3 in the same time interval.

(2) *Rhythmicity.* The pulse is rhythmic when all subsequent time intervals τ have the same duration, arrhythmic when subsequent τ differ from each other (as in the atrial fibrillation).

(3) *Size*: depends on the *cardiac output pulse*: i.e. on the amount of blood ejected by the left ventricle at each systole; e.g. if instead of 60 ml of blood, the left ventricle injects 120 ml of blood the *size* of the sphygmic wave will be double, provided that the resistance upstream the location where the pulse is measured is the same.

(4) *Hardness.* Under our finger we can feel a hard cord requiring some pressure to be squeezed or a soft structure. The hardness is greater the more rigid are the walls of the vessel (as in the arteriosclerosis) or the higher is the blood pressure (as in hypertension).

(5) *Rapidity:* dP/dt. For example in the arteriosclerosis (Fig. 1.22c) or in the aortic insufficiency (Fig. 1.22d) the pulse is more *rapid* than in the stenosis of the aorta (Fig. 1.22b).

1.5.1 *Measurement of Blood Pressure*

As shown in Fig. 1.19b, blood pressure can be measured *directly* by means of catheters, provided that the opening of the catheter does not interfere with blood velocity (middle picture in Fig. 1.19b). More common however are *indirect* methods of measurement with two methods the *sphygmomanometer of Riva Rocci* and the *oscillometer of Pachon*. Figure 1.23 shows a schema of the *sphygmomanometer of Riva Rocci*: an armlet is bound over the upper segment of the arm so that its height is about similar to that of the heart (see changes in pressure with height in the Static section of these notes) and a phonendoscope is placed just below its lower border. The armlet is made up by an inextensible external band and by an extensible internal band. When air is pumped into the armlet, the internal band compresses the arm and the humeral artery within it. When the applied pressure is greater than the maximal systolic pressure the artery is closed and no noise will be heard with the phonendoscope. Air is then let escape slowly through the exhaust. As soon as the pressure is *just lower* than the maximal blood pressure a small noise will be heard for a short fraction of the period τ, i.e. the short time interval during which blood can flow below the armlet with a noisy turbulent motion: the pressure in this interval corresponds to the *maximal pressure* (or just below it). When the pressure in the armlet is allowed to decrease below this value both the duration and the noise increase reaching a maximum at the level of the *average pressure*: in this case the artery will be open for about one half of the period τ. When pressure is allowed to fall further, the artery will remain open for most of τ and closed for a short fraction of it; the noise heard in

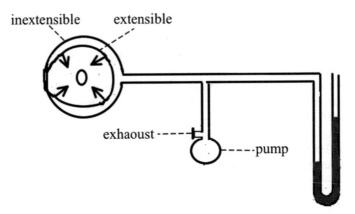

inextensible   extensible

exhaoust ----

----pump

Fig. 1.23 A schema of the *sphygmomanometer of Riva Rocci*. An armlet made up by an external inextensible and by an internal extensible material is bound over the upper arm. The pressure indicated by the arrows compressing the humeral artery is obtained by pumping air within the armlet and is measured by means of a manometer filled with mercury. After a maximum is attained, sufficient to close the artery, air is allowed to escape slowly through the exhaust

this case occurs when the artery will close for a short time interval after its full open condition: this short *neat noise* corresponds to the *minimal pressure*. If the pressure is allowed to fall further the artery will not close completely, but its section will oscillate from a maximum to a minimum value where turbulent motion may also occur resulting in a lower, *muffled noise*, i.e. a sound wave with harmonics of lower frequency above it. It is therefore inexact to take as minimal pressure the end of the noise, but rather the change of the noise from neat to muffled.

In order to *record* instead of hear the oscillations of the artery described above it is used the *oscillometer of Pachon*. Small oscillations of the column of mercury within the manometer can also be viewed when the pressure in the armlet equals the mean blood pressure (i.e. when the oscillations of the artery attain a maximum), but these oscillations are barely visible because the oscillating mass (the column of mercury) is large. We need a thin, sensitive structure to record the oscillations. The problem however is that this thin structure will blow up under the high pressure in the armlet. This problem is solved by the *oscillometer of Pachon* (Fig. 1.24).

The thin, delicate and sensitive capsule is enclosed in a rigid box and a communication is allowed between its inside and the box around it. When air is pumped in the three cavities (armlet, capsule and box) nothing happens because the pressure is equal inside and outside the sensitive capsule. However if after pumping air within the system the communication L between capsule and box is closed, the oscillations of the pressure will be transmitted to the capsule only and not to its outside. The surface of the capsule will then oscillate recording the small pressure changes taking place over the large background pressure over which these changes take place (Fig. 1.25).

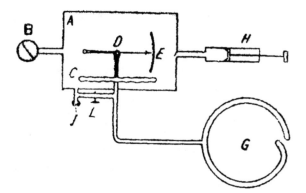

Fig. 1.24 Oscillometer of Pachon: *G* armlet; *H* pump; *B* manometer; *C* sensitive aneroid capsule within the rigid box *A* and attached to an index *D* and scale *E*; *L*: communication between inside of *C* and *A*; *J* exhaust of air (from Margaria R and De Caro L *Principi di Fisiologia Umana*, Francesco Vallardi, Milano, 1977)

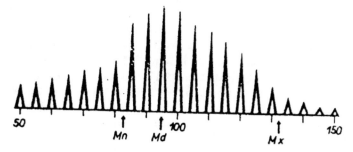

Fig. 1.25 Amplitude of the oscillations recorded with the Oscillometer of Pachon as a function of the background pressure in mm of mercury over which they take place. *Mx, Md* and *Mn* indicate respectively the maximal, average and minimal blood pressure (from Margaria R and De Caro L *Principi di Fisiologia Umana*, Francesco Vallardi, Milano, Italy, 1977)

As it can be seen from Fig. 1.25 the *average* pressure is clearly indicated by the highest peak attained by the oscillations; less precise is the measurement of the maximal *Mx* and the minimal pressure *Mn*, which, as described above, can be better determined from the first noise and the last *neat* noise heard with a phonendoscope when the pressure within the armlet is allowed to fall slowly.

Why the maximal oscillation in Fig. 1.25 takes place when the pressure within the armlet equals the *average blood pressure*? If the time-average pressure in the armlet is greater than the time-average blood pressure, the wall of the artery will be charged from outside to inside. The contrary is true when the pressure is less than the average: in this case the walls of the artery will be charged from inside to outside. Only when the time-average pressure in the armlet equals the time-average pressure during τ the walls of the artery will be discharged and will respond to the pulse wave with a maximal oscillation.

1.5.2   Organization of Systemic and Pulmonary Circulation

In Fig. 1.1 the systemic circulation is divided into two parts: one part at high pressure on the right (arteries) and one part of low pressure on the left (veins). The two parts are in communication through paths that are in parallel with each other. This representation is interesting because it shows that this organization allows to modify the blood flow in one district without interfere appreciably with blood flow in the other districts. For example blood flow increases in the digestive tract after dinner and in muscles during running independently of each other (except in pathological conditions as in the anaphylactic shock). On the contrary if the districts were in series, as in the kidney, an increase in blood flow in the glomeruli will increase, *ceteris paribus*, the blood flow in the tubules.

The crosses in Fig. 1.1 represent the *peripheral resistance units*, PRU, which are represented by the arterioles. The arterioles represent the *greater* and the *more variable* resistance of the circulation.

As described above, in the *Law of Poiseuille*: $\Delta P = (8l\eta/\pi R^4)$ *Flow*, the term $(8l\eta/\pi R^4) = \Delta P/Flow$ is the *Resistance* offered to the flow of the fluid into the vessel. The term l/R^4 is called the *geometric* factor of the resistance.

The units of the PRU $= \Delta P/Flow$ are usually expressed as ΔP in mm of mercury (Hg) divided by the *Flow* in ml/s. This is convenient because in normal conditions in the systemic circulation, the PRU expressed in that way *approach unity*. In fact, at rest, blood *Flow* is 5 l/min = 5000 ml/60 s = 83 ml/s and $\Delta P = 100-2 = 98$ mm Hg; i.e. PRU = 98/83 = 1.2. For example suppose that in hypertension $\Delta P = 160$, then PRU = 160/83 = 1.9. In an anemic subject with a lower concentration in blood cells the viscosity η is reduced and, given a constant $\Delta P = 100-2 = 98$ mm Hg, the blood flow will increase due to the smaller resistance, e.g. to 100 ml/s, and PRU = 98/100 = 0.98. Let's consider the muscular exercise: the flow increases, e.g. three times, i.e. $83 \times 3 = 249$ ml/s, also ΔP increases but not as much as the flow, suppose to 140 mm Hg, then PRU = 140/249 = 0.56.

In stationary conditions the blood output of the left ventricle *must* equal the blood output of the right ventricle. It would be very difficult for an engineer to construct two pumps with the *exact* same output. Fortunately, as we will see later, Nature provided a mechanism (the Starling law) capable to attain this goal. Figure 1.26 shows that dead would quickly ensue (~10 min) if the mean output of the right ventricle into the lungs were only 2% greater than that of the left ventricle in the systemic circulation. This is because the congested pulmonary capillaries will emit liquid into the alveoli preventing gas exchange: foam will fill up the airways in this condition.

Since, as described above, the blood *Flow* is the same in each total section of the vessel tree (Fig. 1.18) and since the resistance PRU $= \Delta P/Flow$, the pressure changes ΔP are proportional to changes in *resistance* offered by each district. From Fig. 1.27, it can be seen that: (*i*) the pressure (i.e. the resistance) in the pulmonary circuit (right) is about 1/6 that on the systemic circulation (left); (*ii*) the sphygmic wave is progressively damped from center to periphery and is nil in the capillaries of the systemic circulation where the pressure falls from 35 to 15 mm Hg with an

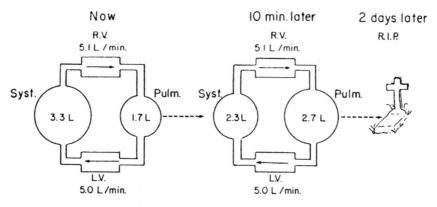

Fig. 1.26 Left: initial condition: 3.3 l of blood are contained in the vessels of the systemic circulation (*Syst.*) and 1.7 l in those of the lungs (*Pulm.*). Right: in case the right ventricle of the heart (R.V.) would exceed that of the left ventricle (L.V.) of only 100 ml/min, dead would ensue in 10 min due to pulmonary congestion (from Burton AC *Physiology and Biophysics of the circulation*, Year Book Medical Publishers Incorporated, Chicago, 1965)

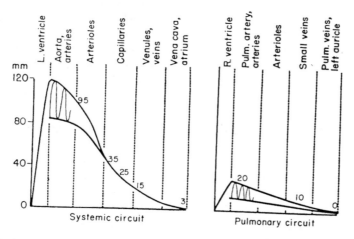

Fig. 1.27 On the abscissae are indicated the different kind of vessels; on the ordinate the progressive fall in pressure from arteries to veins (from Burton AC *Physiology and Biophysics of the circulation*, Year Book Medical Publishers Incorporated, Chicago, 1965)

average of 25 mm Hg (figures that, as we will see later, should be memorized); (*iii*) the fall in pressure, i.e. the resistance offered, is largest in the arterioles district: about 60 mm Hg; as described above, this large resistance is also the most variable allowing a proper distribution of blood flow between the different parallel ways connecting the high pressure to the low pressure side of the systemic circulation (Fig. 1.1).

The fall in pressure in the systemic circulation is greatest at the level of the arterioles in spite of the fact that their total cross-sectional area is much greater than that of the aorta. This can be qualitatively understood considering that the division of a vessel of area πR^2 into two branches having the same total cross-sectional area $\pi \left(r_1^2 + r_2^2\right)$, doubles the surface where the gradient of velocity is higher and, consequently, doubles the frictional losses between blood layers. In addition, the *Flow* in the first conduct equals the sum of the flows in the two branches having the same total cross-sectional area. According to the *Law of Poiseuille*:

$$\Delta P_1(\pi R^4/8l\eta) = \Delta P_2(\pi r_1^4/8l\eta) + \Delta P_2(\pi r_2^4/8l\eta)$$
$$\Delta P_1 R^4 = \Delta P_2(r_1^4 + r_2^4)$$

i.e. $\Delta P_1 < \Delta P_2$ because $R^4 > \left(r_1^4 + r_2^4\right)$ since $R^2 = \left(r_1^2 + r_2^2\right)$.

Figure 1.28 shows a way to measure the velocity of the sphygmic wave. The propagation of a wave is not a transport of matter, but a transport of energy (consider e.g. the sound waves). If the arteries were perfectly rigid the velocity of propagation of the sphygmic wave would approach that of sound in water. However the walls of the arteries are extensible and elastic, more exactly *visco-elastic* (similar to an elastic spring in parallel with a dashpot). This implies that work must be done during propagation not only to transmit kinetic energy between liquid molecules but also to deform, against friction, the walls of the blood vessels. This work requires time to be done and, as a consequence, a slowdown of the propagation of the wave. In addition some energy will not be stored elastically during the expansion of the walls of the vessel, but degraded into heat by friction with the consequence that the amplitude of the wave will decrease along its path. It follows that: (*i*) the velocity of the sphygmic wave will be greater the more rigid are the walls of the vessel, and (*ii*) its amplitude will decrease from center to periphery of the vascular bed.

What is the velocity of propagation of the sphygmic wave? An *empirical* equation (by Bramwell JC and Hill AV, *Lancet*, 1922) is: v (m/s) = 3.6/$\sqrt{\text{Extensibility}}$, where the Extensibility (or compliance: $\Delta V/\Delta P$) is expressed as % of initial volume/1 mm Hg. Reasonable values of v are 5 m/s in a 5 years old kid and 8.5 m/s in a 85 year old man; in a normal adult $v = 6$–7 m/s. Since the stiffness of blood vessels $\Delta P/\Delta V$ increases with the transmural pressure P (Figs. 1.20b, 1.31a), a progressive deformation of the sphygmic wave takes place along the arterial tree due to the fact that its less extensible upper part (P_{max}) travels faster than the more extensible lower part (P_{min}) (as mentioned above this is similar to the height of sea waves approaching shore).

How long is the tract of the arterial tree interested by the increase in pressure ΔP? Since the velocity (7 m/s) is $v = d/t$, then $d = vt$. Assuming that the increase in pressure during the sphygmic wave lasts ~0.3 s (see Fig. 1.21) ΔP will invade a tract of the arterial tree of $d = 7 \times 0.3 = 2.1$ m. This means that the *entire* arterial tree will expand during the systole.

Now let's consider the *time of circulation*, t_c i.e. the time a given mass of blood requires to return to its initial position. Since *Flow* = *Volume/time*, then $t_c = $ *Volume/Flow*. Since the *Flow* is equal through all sections of the systemic tree, t_c will be greater the greater the volume of the district in which blood flows, which is rep-

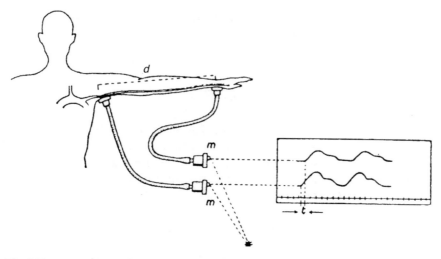

Fig. 1.28 Measurement of the velocity of the sphygmic wave $v = d/t$. Two metallic capsule closed by a rubber membrane and connected by two optical manometers m are used to measure the time difference t between two sphygmic waves at the distance d (from Margaria R and De Caro L *Principi di Fisiologia Umana*, Francesco Vallardi, Milano, Italy, 1977)

resented by the *venous* bed (about 3 l of the 5 l of the total volume). Measuring $t_c = 5 \text{ l}/(5 \text{ l/min}) = 1$ min is nonsense because t_c differs in the different districts. A way to approach the total value of t_c is to inject in a vein of the arm a bitter substance and to measure the time delay between injection and a bitter taste in the mouth; in this case the time required for the substance to travel from the arm vein, to the venous bed, cava veins, right heart, pulmonary circulation, left ventricle, arteries, arterioles and capillaries of the mouth is about 30–40 s. This is a minimum value because it is short of the time for the substance to go from the capillaries to the tract of veins uphill the location of injection.

1.6 Equilibrium Conditions of Blood Vessels

The *pressure* $P = Force/Surface = Fl^{-2}$ across the walls of a vessel causes a *tension* $T = Force/Length = Fl^{-1}$ to which the walls are subjected. To understand the difference between P and T remember when as a kid a balloon was inflated for you: when the small thick-walled balloon was inflated by pumping air with pressure P within it, you observed that the walls of the balloon became transparent; this is because the layers of rubber did slide one over the other; the force T parallel to the surface and perpendicular to the unit of surface length is the cause of this sliding; if the pressure increases too much the connections between surface layers is lost and the balloon explodes because *tension* slides surface layers apart; *tension* not *pressure* is the ultimate cause of rupture of an organ. Therefore in the blood vessels rupture is caused by the *tension* not by the *pressure*.

1.6.1   Tension Elastic (Passive), Active and Mixed

The tension on the walls of the vessel caused by blood pressure inside the vessel is called *passive or elastic tension*. However consider the heart: the active force of the contracting muscular fibers *acting parallel to the surface* is an *active tension* that causes the pressure of the blood inside the heart cavities. In both cases the relationship between pressure P and tension T is established by the *Law of Laplace*:

$$P = (1/R_1 \pm 1/R_2)T$$

where R_1 and R_2 are the rays of curvature of two arcs on the surface *perpendicular* to each other. The rays of the two arcs that differ more from each other are called *principal rays* of curvature. For example, consider a *straight* cylinder, the principal rays of curvature are: (*i*) infinity ∞, along the length of the tube and (*ii*) equal to the radius R of the tube along its conference. The Laplace law for a straight *cylinder* will then be

$$P = (1/\infty + 1/R)T = T/R$$

In the *sphere* the rays of curvature R of two arcs perpendicular to each other are the same in each point of the surface and

$$P = (1/R + 1/R)T = 2T/R$$

Now let's imagine to bend a cylindrical tube; in this case we will have two opposite surfaces of the tube: an upper surface where the two principal rays of curvature have their center on the same side (below the surface) and a lower surface where the two principal rays of curvature have their center in opposite sides (one below and the other above the surface). In the first case the surface is called *synclastic* and the two terms of the *Law of Laplace* add:

$$P = (1/R_1 + 1/R_2)T$$

In the second case the surface is called *anticlastic* and the two terms subtract:

$$P = (1/R_1 - 1/R_2)T$$

It follows that, at a given pressure P inside the tube, the tension T is lower in the synclastic surface and greater in the anticlastic surface. An example is the *arch of*

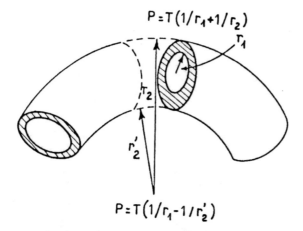

Fig. 1.29 Arch of the aorta where the upper wall is synclastic and the lower wall is anticlastic. As explained in the text the thickness of the walls changes correspondingly (from Burton AC *Physical principles of circulatory phenomena: the physical equilibria of the heart and blood vessels*, in Hamilton WF and Dow P (eds) *Handbook of Physiology*, Sect. 2, Vol. 1, Am. Phys. Soc., Washington DC, 1962)

the aorta: the elastic tension is lower in the upper synclastic surface and higher in the lower anticlastic surface. In fact the thickness of the wall is greater in the lower than in the upper surface, i.e. is greater where is greater the elastic tension caused by blood pressure (Fig. 1.29).

Figure 1.30 shows the tension on the walls that we have discussed till now, i.e. the *circumferential tension*: $T_c = PR$. In the case of a cylinder, as in Fig. 1.30, this tension tends to divaricate the edges of a small cut parallel to the longitudinal axis of the vessel.

However the pressure P acting on the whole circulatory tree stretches the vessel also longitudinally resulting in a *longitudinal tension* T_1 caused by a force acting on two imaginary diaphragms of area πR^2 that close the ends of the vessel. This tension tends to divaricate the edges of a small cut perpendicular to the longitudinal axis of the vessel. In reality two diaphragms do not exist, but it is intuitively understandable that the ramifications of the circulatory tree act as two diaphragms. The force stretching longitudinally the vessel is $F = P\pi R^2$. This force acts perpendicularly to the circumference causing the longitudinal tension:

$$T_l = P\pi R^2 / 2\pi R = PR/2$$

We have seen that the circumferential tension in a cylinder according to the Law of Laplace is $T_c = PR$; it follows that the longitudinal tension in a cylindrical vessel is one half the circumferential tension. From now on we will always consider the circumferential tension.

Fig. 1.30 Circumferential
tension T caused by the
pressure P in a cylinder of
radius R (from Burton AC
*Physical principles of
circulatory phenomena: the
physical equilibria of the
heart and blood vessels*, in
Hamilton WF and Dow P
(eds) *Handbook of
Physiology*, Sect. 2, Vol. 1,
Am. Phys. Soc., Washington
DC, 1962)

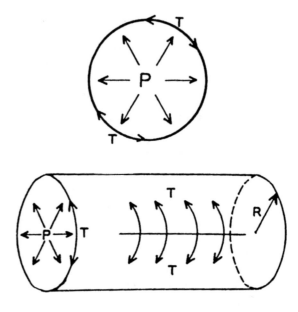

As mentioned above there are two kind of tension: *passive or elastic* and *active*.
The *passive* tension: (*i*) is fully dependent on the *transmural* pressure P distending the
walls of the vessel, (*ii*) does not require energy expenditure and (*iii*) confers stability
to the vessel. In contrast the *active* tension: (*i*) can be adjusted independent of the
transmural pressure, (*ii*) requires energy expenditure and (*iii*) confers *instability* to
the vessel.

Now, let's consider the *passive-elastic tension* in the wall of the aorta having the
radius of ~1 cm, and compare it with that on the wall of the capillaries having a
radius about equal that of a red blood cell ~4 μm (4×10^{-4} cm). Using the CGS
system (remember that 1 mm Hg corresponds to 1300 dyne/cm^{-2} as described in
Sect. 1.3.1):

$$T_{aorta} = PR = 100 \text{ mm Hg} \times 1300 \text{ dyne cm}^{-2}/\text{mm Hg} \times 1 \text{ cm} = 130,000 \text{ dyne/cm}$$

$$T_{capillaries} = PR = 25 \text{ mm Hg} \times 1300 \text{ dyne cm}^{-2}/\text{mm Hg} \times 4 \times 10^{-4} \text{ cm} = 13 \text{ dyne/cm}$$

The above comparison shows that in spite of the fact that the pressure in the
capillaries is only one fourth that in the aorta the tension in their walls is ten thousand
times smaller due to the reduction of their size! A tension of 13 dyne/cm is *very* small.
Imagine that the tension at the interface air-water over which some insects 'walk'
(Fig. 3.22) is 72 dyne/cm! This example stresses that the resistance offered by the
walls of a blood vessel to yield under its transmural pressure is lower the greater its
radius of curvature (consider an aneurism); the smaller the radius the more resistant
is the vessel. Accordingly the amount of elastic tissue is maximal in the walls of the
aorta and in the large arteries and nil in the capillaries where the exchanges take place
through their walls. In the veins the pressure falls, but the radius increases and with

Fig. 1.31 **a** Pressure
(abscissa)–volume (ordinate)
relations of aorta and vena
cava segments. Note the
flattening of the cava vein at
low transmural pressure. **b**
Relations between tension on
the walls (ordinate) and
radius of the vessel
(abscissa) calculated as
described in the text from the
pressure-volume relations in
(**a**) (modified from Burton
AC, in Ruch TC and Patton
HD (eds): *Physiology and
Biophysics*, W.B. Saunders
Company, 1966)

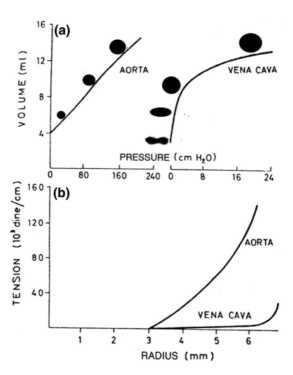

it the tension: in the cava vein pressure is only 10 mm Hg, but the radius increases to
1.6 cm resulting in a tension of 21,000 dyne/cm and in an increase in elastic tissue
in the walls. In the arterioles the tension is about 60,000 dyne/cm and is sustained
by passive elastic structures *mixed* with contractile muscle cells, which as described
above allow a variable resistance to different circulatory districts (Fig. 1.1).

If only elastic structures were subjected to a transmural pressure (as a rubber
balloon), the elastic elements may slide one over the other up to a limit where
connection between them is lost and the organ will break. This problem, common to
all organic structures of our body (vessels, hearth, lungs, muscles), is solved in three
ways: (*i*) by surrounding the elastic structure with *inextensible* elements (usually a
collagen net) opposing an excessive increase in volume or length; (*ii*) the progressive
recruitment with increasing volume of parallel elastic structures of increasing length
in the wall of the organ resulting in a progressive increase in its stiffness; (*iii*) the
nonlinear force-length relationship of the single elastin fiber. As a consequence of
these three mechanisms the *pressure-volume diagram* of a hollow organ is not linear,
but shows a progressive increase in pressure ΔP for a given increase in volume ΔV.
More appropriate to investigate the stress at which the walls of the organ are subjected
is a *tension-radius diagram T-R*, which can easily obtained from the *P-V* relation for
a cylindrical vessel of length *l* since $T = PR$ and $R = \sqrt{(V/\pi l)}$ (Fig. 1.31).

The thick curve in Fig. 1.32 indicates the *elastic* tension *T* that *actually exists
at a given radius R* depending on the *anatomy* of the vessel. The thin straight lines

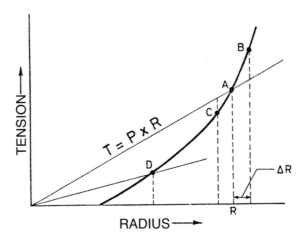

Fig. 1.32 The thick line shows the relationship existing between the radius of blood vessel (abscissa) and the *elastic* tension of its walls (ordinate): the curve has been obtained as described in Fig. 1.31. The two thinner straight lines show the relationship that must exist between tension and radius according to the law of Laplace: the slope of the straight lines indicates the difference in pressure P across the walls of the vessel. Equilibrium between tension caused by the pressure P and the elastic tension of the walls takes place in the point of intersection between straight thin lines and the thick curve. If the slope P of the thin straight line decreases, equilibrium will be attained at a lower radius D. As described below, when *elastic* tension only acts on the vessel walls the equilibrium is stable (from Burton AC, *Amer. J. Physiol.* 164:319, 1951)

$T = PR$ indicate the tension T that *must exist* according to the law of Laplace at a *given pressure P and radius R*. It follows that in equilibrium conditions the radius of the vessel will correspond to the point of intersection between thin and thick lines (points A and D in Fig. 1.32). The slope of the thin straight lines is the pressure P: point A indicate a point of intersection at a radius greater than at point D because the pressure P (the slope of the straight line) is greater.

The *elastic* tension, increasing with the radius of the vessel, confers *stability* to the vessel. Consider the equilibrium point A in Fig. 1.32. An accidental increase ΔR from the equilibrium position will result in an increase of the elastic tension to point B, i.e. to a value greater than that required by the law of Laplace for that pressure. It follows a tendency of the radius to return to the point A. The contrary is true in case of an accidental decrease in radius to point C. The equilibrium condition is that of a ball in the bottom of a cup: ‿•‿.

If the elastic diagram were not concave towards the tension axis, but were rectilinear or concave towards the radius axis (Fig. 1.33), equilibrium would be possible only at low pressure values and an increment in pressure P would easily cause a rupture of the vessel.

In Fig. 1.34 the thick line represents the elastic tension-radius of a vessel whose walls are affected by a lesion: this causes an inflection of the elastic diagram. An increase in pressure from P_1 to P_2 causes a small increase of the radius from R_1 to

Fig. 1.33 Hypothetical case
in which the elastic
tension-radius diagram (thick
curve) were concave towards
the radius axis. Equilibrium
would be possible only for
pressure values smaller than
P_2 (from Catton WT,
*Physical methods in
physiology*. Sir Isaac Pitman
& Sons, Ltd. London, 1957)

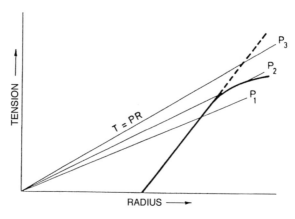

Fig. 1.34 Elastic
tension-radius diagram (thick
line) of a vessel whose walls
are weakened by illness. A
small increase in pressure
from P_2 to P_3 causes a
relevant increase in the
radius of the vessel (from R_2
to R_3) (from Burton AC, *Am.
J. Physiol.* 164: 319, 1951)

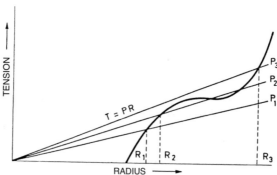

R_2, but the *same* increment of pressure rom P_2 to P_3 causes a sudden large increase
in radius from R_2 to R_3: this is how an *aneurism* occurs.

1.6.2   Critical Closing Tension and Pressure

The *active* tension, depending on the frequency of the action potentials on the muscular fibers in the vessel walls, is *independent of its radius*: this is indicated by the
horizontal thick line in Fig. 1.35. As described below the active tension confers
instability to the vessel.

In fact, an accidental increase of the radius ΔR from the equilibrium position in
A will require a tension B greater than that at disposal with the consequence that
the radius will increase further until full open. The contrary is true in case of an
accidental decrease in radius to C: in this case the active tension on the vessel walls
will exceed that opposed according to the law of Laplace by the transmural pressure
and the vessel will collapse. The equilibrium in A of Fig. 1.35 is unstable like that of

Fig. 1.35 Unstable
equilibrium in case of an
active tension independent of
the radius of the vessel (from
Burton AC, *Am. J. Physiol.*
164: 319, 1951)

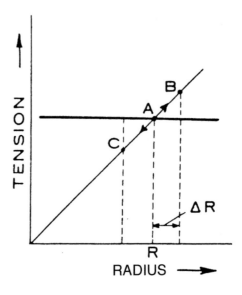

a ball on top of an upside-down cup: ⌢. This condition applies to some bladder
and intestinal sphincters.

Actually in the vessels (e.g. in the arterioles) both elastic and muscular fibers
coexist creating a *mixed* tension. This is the most general case, which more commonly
applies to blood vessels (Fig. 1.36). Suppose that at the beginning the muscular fibers
are relaxed and to have a stable equilibrium at the point A between elastic tension
$(A - R_2)$ and tension $T = PR$ caused, according the Laplace law, by the pressure P
(slope of the thin straight line). If now an active tension T_A $(B\text{-}C)$ adds to the elastic
tension, the radius will decrease (and with it the elastic tension) from R_2 to R_1 until
the total tension of the walls $T = T_A + T_E$ equals that required by the law of Laplace
for that pressure (point B on the thin straight line). If the active tension increases
further, suppose to $(D\text{-}E)$, the radius of the vessel, and with it the elastic tension,
will decrease until room is made for the active tension between thin and thick lines;
an additional small increase of the active tension will result in a complete closure
of the vessel. The maximal distance between the elastic tension (thick line) and the
tension caused by the transmural pressure P (thin line) is called *active tension critical
of closure*. Similarly if the active tension T_A is maintained constant the vessel will
collapse if the pressure (i.e. the slope of the thin line) decreases in such a way that
the total tension of the walls (elastic + active) is greater than $T = PR$ for any value
of R: the minimal pressure below which closure of the vessel takes place (given a
constant active tension) is called *critical pressure of closure*. The critical pressure of
closure will be obviously greater the greater the active tension, i.e. the greater the
vasomotor tone. Note that closure of a vessel may occur downstream the point where
a constriction of the vessel (e.g. an atheromatous plaque) is the initial cause of the
fall in pressure.

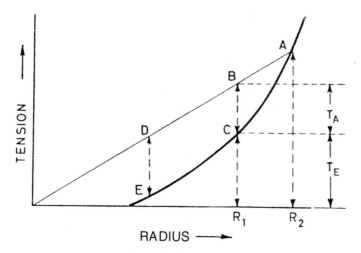

Fig. 1.36 Equilibrium conditions in case the walls of the vessel were simultaneously subjected to a passive elastic tension T_E and to a muscular active tension T_A (from Burton AC, *Am. J. Physiol.* 164: 319, 1951)

In conclusion: the elastic tension provides stability to blood vessel by allowing an automatic regulation of its radius accordingly to the changes in blood pressure. The active tension alone, which on the other hand is required to regulate a proper distribution of blood, makes the vessel completely unstable. In the circulatory tree a compromise is attained by associating elastic and active tensions: the vessels with walls predominantly elastic are more stable and less sensitive to changes of the vasomotor tone; the contrary is true for the vessels with predominantly muscular walls (as the arterioles). Figure 1.36 shows that the change of the radius of a vessel following an even slight increase of the active tension T_A is strictly bound to the average blood pressure: when this is low (as in the *hypotensive shock*) an increase, even small, of the active tension may lead to complete closure of the vessel.

What described above, explains the trend of the flow-pressure curves (Fig. 1.37), which differ appreciably from the linear relation predicted by the law of Poiseuille:

$$\Delta P = (8l\eta/\pi R^4)Flow = Resistance \times Flow$$

from which it can be seen that, for a given viscosity η and geometry of the vessel, the *Resistance*, R, is constant for any value of ΔP: i.e. $R = \Delta P/Flow = $ constant. From Fig. 1.37 on the contrary it can be seen that R is constant only at high values of pressure. This is due to the finding that the pressure-flow curves do not start from the origin (as predicted by the law of Poiseuille), but from a finite value of pressure: the *critical pressure of closure* ΔP_c, i.e. $\Delta P = \Delta P_c + b$ $Flow$ and $R = \Delta P_c/Flow + b$ from which it can be seen that when the flow tends to zero R tends to infinity as indicated in the right hand graph of Fig. 1.37 (this is because ΔP_c is constant independent of the flow, contrary to the Poiseuille law where ΔP is zero when the

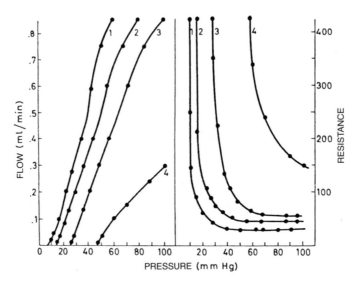

Fig. 1.37 Flow-pressure curves determined on the vessels of the ear of a rabbit (left) and resistance-pressure curves (right) calculated from the left–hand curves (resistance = pressure/flow). According to the law of Poiseuille, the flow-pressure relationship would be a straight line starting from the origin and the reciprocal of its slope, the resistance, would be constant for each value of pressure (the right–hand graph would be an horizontal line). Curves from 1 to 4 were obtained by increasing, progressively, the vasomotor tone of the blood vessels through an increasing stimulation of the sympathetic nerves; this results in an increase in the critical closing pressure and in a deviation from the law of Poiseuille (from Burton AC, in Ruch TC and Patton HD (eds): *Physiology and Biophysics*, W.B. Saunders Company, 1966)

flow is zero). ΔP_c is greater the greater the vasomotor tone. In case the vasomotor tone is small it can be seen that resistance changes little with increasing pressure for a large range of pressure. The critical pressure of closure is therefore the principal cause of deviation from the law of Poiseuille of the pressure-flow relation in the blood vessels.

1.7   The Capillaries

1.7.1   Function of Capillaries

The capillaries are the ultimate goal of both systemic and pulmonary circulations. This is because, as described at the beginning of these notes, requirement of life is to maintain constant the physics and the chemistry of the organism; this function implies exchanges between inside and outside of the body which ultimately take place at the level of the capillaries. The exchanges are favored by the following characteristics of capillary circulation.

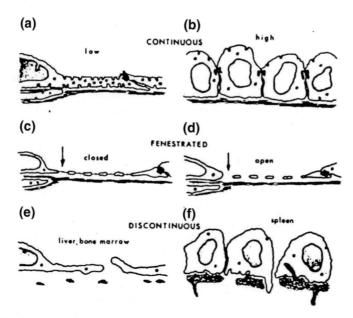

Fig. 1.38 Classification of capillaries walls (from Majno G, *Ultrastructure of the vascular membrane*, Hamilton WF and Dow P (eds) *Handbook of Physiology*, Vol. 3, Sect. 2, American Physiological Society, Washington, D.C., 1965)

(1) As shown in Fig. 1.18 the same blood flow in the aorta is distributed in the larger comprehensive sectional area of all capillaries (~700 times that of aorta); this results in a proportional decrease in blood velocity in the capillaries (velocity = flow/section) and, as a consequence, in an increase of the time at disposal for the exchanges to occur while the blood travels the capillary length (~1 mm): about 2.5 s for a cardiac output at rest of 5 l/min.

(2) The partition of a large vessel (e.g. aorta) in a multitude of smaller vessels (the capillaries) increases the overall surface where the exchanges can occur.

(3) As described in Sect. 1.6 the walls of capillaries are subjected to a very small tension (~13 dyne/cm) sustained by thin unicellular structures offering little resistance to the exchanges between inside and outside of the vessel (Fig. 1.38).

(4) The diameter of the capillaries about equals that of an erythrocyte (~8 μm), which changes its shape by contact with the capillary walls allowing mixing of its hemoglobin content and a more uniform exchange of oxygen and carbon dioxide with the outside surrounding.

1.7.2 *Production of the Interstitial Fluid and Edema*

A liquid called *interstitial fluid* surrounds the capillaries. The composition of this liquid differs from the *plasma fluid* contained in the capillaries and surrounding the blood cells. Exchanges take place between the two liquids, external and internal. These exchanges are governed by two different kind of pressure: the *hydraulics pressure*, that we have considered till now, and the *osmotic pressure*, which depends from a difference in *solute* concentration in two adjacent departments. The *osmotic* pressure equals the pressure required to prevent the passage of water molecules from the department of lower solute concentration to the department of higher solute concentration, through *pores* small enough to allow passage of molecules of water and not of the solute. In our specific case the difference in osmotic pressure is due to the fact that the plasma has a greater concentration of *protein* molecules than the interstitial fluid. The osmotic pressure due to this difference in protein concentration is called *oncotic pressure* and amounts to 25 mm Hg: this means that molecules of water will tend to penetrate inside of the capillaries pushed by a pressure of 25 mm Hg. Since the hydraulic pressure in the capillary of the *systemic circulation* is 35 mm Hg at the arterial extremity and 15 mm Hg at its venous extremity, water molecules will exit pushed by a difference in pressure of 35 mm Hg (hydraulic) − 25 mm Hg (osmotic) = 10 mm Hg at the arterial extremity and reabsorbed by 15 mm Hg (hydraulic) − 25 (osmotic) = − 10 mm Hg at the venous extremity. In other words a circulation of water will take place outside the capillary from its beginning and to its end (Fig. 1.39).

In the small (*pulmonary*) circulation the hydraulic pressure in the capillaries is on the average 10 mm Hg only with the consequence that water absorption takes place along the whole capillary length driven by a pressure difference of 10 − 25 = − 15 mm Hg. In other words the capillaries of the pulmonary circulation act as *suction pumps* preventing a liquid layer over the surface of the alveoli hindering diffusion of gasses between blood and alveolar air. As shown in Fig. 1.26 an increase in hydraulic pressure in the pulmonary capillaries would result into a pulmonary *edema* impeding gas exchanges.

In the above simplified description it was implicitly assumed that the hydraulic pressure of the interstitial fluid outside the capillaries was zero, i.e. that 25 mm Hg − 0 mm Hg = 25 mm Hg were the average gradient of hydraulic pressure between inside and outside of the capillaries in the systemic circulation. Furthermore it was assumed that the concentration of the proteins in the interstitial fluid is nil. More specifically we must consider separately: (1) the hydraulic pressure in the capillaries, $P_{capillary}$; (2) the hydraulic pressure in the tissues, P_t; (3) the *lymphatic* vessels; (4) the oncotic pressure in the capillaries Π_c; (5) the oncotic pressure in the tissues Π_t; (6) the *permeability* of the capillaries. Let's consider these six factors separately.

(1) The flow of blood that enters the capillary $\dot{V}_{in}$ equals that coming out $\dot{V}_{out}$. Since, according to the law of Poiseuille $\Delta P = \text{Resistance} (R) \times \dot{V}$, the flow input into the capillary will be

Fig. 1.39 Systemic circulation: relationship between water inlet in capillary due to the *oncotic* pressure of proteins (O.P. = 25 mm Hg within the whole capillary extent) and water transfer through the capillary due to the *hydraulic* pressure falling from 35 mm Hg at its arterial extremity to 15 mm Hg at it venous extremity; this result in a fluid circulation outside the capillary (round arrow at the left of the figure). The excess of interstitial fluid is adsorbed by lymphatic terminals (right) (from Wright S *Proceedings of the Royal Society of Medicine*, 32: 651, 1939)

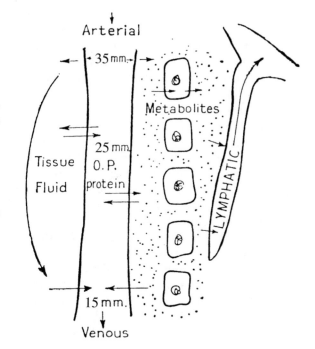

$$\dot{V}_{in} = \Delta P/R_{art} = P_{art} - P_{capillary}/R_{art}$$

where P_{art} is the arterial pressure, R_{art} is the resistance upstream the capillary (PRU) and $P_{capillary}$ is the pressure at the center of the capillary.

$$\dot{V}_{out} = \Delta P/R_{ven} = P_{capillary} - P_{ven}/R_{ven}$$

where P_{ven} is the venous pressure and R_{ven} is the resistance downstream the capillary.

Since $\dot{V}_{in} = \dot{V}_{out}$:

$$P_{art} - P_{capillary}/R_{art} = P_{capillary} - P_{ven}/R_{ven}, \text{ i.e.}$$
$$P_{art} - P_{capillary}/P_{capillary} - P_{ven} = R_{art}/R_{ven}$$

From the above we can conclude that the *pressure in the capillaries depends on the arterial/venous resistance ratio.* Immediately after exercise we can feel our muscles momentarily swelled: this is because the ratio R_{art}/R_{ven} *decreases* (the resistance of the arterioles, PRU, decreases to let more blood into the capillaries of the working muscles), i.e. $P_{capillary}$ *increases* and liquid is momentarily emitted into the tissues around the capillaries. Another example: an insufficiency of the right ventricle, causing an upstream accumulation of venous blood, corresponds to an increase of R_{ven}, i.e. to a decrease of R_{art}/R_{ven} and to an increase of $P_{capillary}$; this will lead to a net output of liquid from the capillaries into the tissues resulting in *cardiac edema* (particularly evident in the legs: remember, from the Statics, that $P + dgh = $ constant).

(2) The hydraulic pressure in the tissues, the *interstitial pressure* P_t, opposes the output of liquid from the capillaries and the development of an edema. P_t depends on the *tension of the skin* surrounding the organ; e.g. it is greater in a finger than in the suborbital tissue. Examples are the anti-G suits made up by an external inextensible tissue and by an internal elastic and inflatable tissue, which prevent excessive swelling of the vessels of the legs and abdomen when the astronauts are subjects to large positive accelerations. A tendency towards an edema (a *sub-edema* condition) will first observed where the tissues are more extensible, i.e. in the suborbital tissue. Finally P_t is strongly dependent on liquid and proteins absorption by the lymphatic vessels.

(3) The oncotic pressure in the capillaries Π_c, usually 25 mm Hg, depends on the protein intake by the organism. This sometimes is insufficient; in this case the oncotic pressure within the capillaries decreases and with it the absorption of the interstitial fluid (edema from hunger). We all remember the sad images of children undernourished with swollen limbs due to protein deficiency. Since most of the proteins are produced by the liver, another condition of protein deficiency with edema is the epatic insufficiency. Protein loss may also occur in some kidney deseases.

(4) The oncotic pressure in the interstitial liquid Π_t depends on the *anatomy* of the capillaries walls (Fig. 1.38). In the muscles the capillaries walls are continuous, but in spite of this, Π_t amounts to 10–30% relative that in blood; in the intestinal villi to 40–60%; and in the liver, bone marrow and spleen, where the capillary wall is discontinuous, to 80–90%.

(5) The *permeability* of the capillaries walls is increased by some substances, as for example the insect toxins or histamine. In this case proteins may escape from the capillaries into the tissues, the reabsorption of liquid from the tissues into the capillaries will decrease and a more or less extensive edema will take place.

(6) The lymphatic vessels (Fig. 1.39) normally retrieve 2–3 l of interstitial fluid/day containing ~200 g of proteins and take it back to the systemic circulation. If this withdraw is interrupted in some body zones, the hydraulic pressure in the tissues will increase and these zones will inflate. An example is that of a parasite, the Filaria, that, by obstruction of the lymphatic vessels, causes the enlargement of some parts of the body (elephantiasis).

Note that factors 1–4 above are the fundamental basis for the liquid exchanges between capillaries and interstitial fluid, whereas factor 5 affects factor 4, and factor 6 affects factor 2.

Now let's consider two particular cases: the *hypertension* an the *hemorrhage*. In the hypertension the increase in pressure (e.g. up to 160 mm Hg) is confined *upstream the arterioles* (PRU) and, as a consequence, does not affect the capillary-interstitial liquid exchange.

Fig. 1.40 Arrows: displacement of liquid across the capillary membrane. Interrupted line: normal condition. The large vasoconstriction of the arterioles taking place in the hemorrhage causes a fall in the average hydraulic pressure in the capillaries (continuous line). The production of the interstitial fluid stops and liquid is absorbed from the interstitial spaces into the capillary along its whole length. P.O. is the oncotic pressure of the plasmatic proteins (25 mm Hg) (from Green JH, *An Introduction to human physiology*, Oxford University Press, London, 1968)

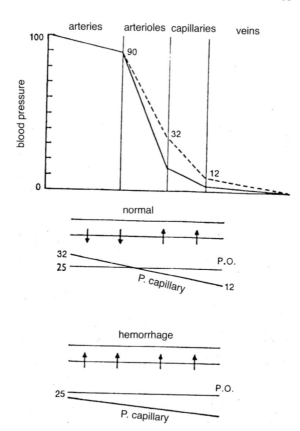

In the hemorrhage the concentration of the blood cells (i.e. the *hematocrit* = volume of blood cells/total blood volume, ~0.45 normally) progressively decreases. Figure 1.40 explains why. The loss of blood during the hemorrhage tends to cause a fall in the arterial pressure below its normal value, e.g. ~90 mm Hg. This is prevented by a constriction of the arterioles (continuous line) with the result that the hydraulic pressure in the capillaries decreases; as a consequence the interstitial fluid, normally produced in the first half of the capillary length and absorbed in the second half (upper case in Fig. 1.40), will be absorbed all along the capillary length (lower case in Fig. 1.40) decreasing the relative concentration of the blood cells, i.e. the hematocrit. Remember that according to the law of Poiseuille ($\Delta P = (8l\eta/\pi R^4) \, Flow$), small variation in the radius R (which is elevated to the *fourth* power) result in very large changes of the resistance ($8l\eta/\pi R^4$) and as a consequence of the pressure change ΔP; small changes of the radius of the arterioles are therefore very effective in the regulation of blood pressure: *the arterioles represent the largest but also the most variable resistance of the circulation.*

Note that 25% of the blood is contained in the small vessels at the periphery (arterioles, capillaries, post capillary venules) and 75% is contained in the large central vessels (heart, large veins and arteries). If the volume of the peripheral vessels increases excessively (e.g. instead of 700 times that of the aorta to 2000 times, as in the *anaphylactic shock*) the blood pressure will fall to values incompatible with life. Another important condition to consider is the *diarrhea in the newborn* where the total blood content may be as low as 200 ml; in this case the loss of liquid, which ultimately derives from the blood, may lead to death in a very short time and it is essential to intervene with a liquid injection as soon as possible.

1.7.3 Venous Circle

The veins are the largest container of blood in our organism. In the adult about 3 l of the total 5 l of blood are contained in the veins. This is because of two characteristics of the veins: (*i*) their larger *extensibility* ($\Delta V/\Delta P$), and (*ii*) their *shape* (Fig. 1.31). The greater extensibility is shown in Fig. 1.31 by the fact that the same increase in volume requires a transmural pressure 10 times greater in the aorta than in the cava vein. The difference in shape is evidenced by the flattening of the cava vein at low-pressure values versus the circular shape maintained by the aorta at all pressure values. This means that in order to increase the volume of the aorta the tension of the walls must increase over the whole pressure range, whereas the flattening of the veins in the low-pressure range allows an increase of the cross-section *without* stretching the walls of the vessel. As shown in Fig. 1.31 this allows an appreciable increase in volume of the venous container with a very low increase in pressure, until a circular section is attained; at this point an inflection of the *P-V* curve takes place and the ratio $\Delta P/\Delta V$ increases even if less than in the aorta due to the greater extensibility of the venous wall. This shows why the venous bed is the largest container where blood stays longer, i.e. the *time of circulation = Volume/Flow* is greater.

The factors that promote the flow of blood into the venous circle towards the heart are: (1) a difference in pressure *P*; in fact, as described previously, motion is sustained by a difference in *mechanical energy* ($E_m/V = dgh + dv^2/2 + P$), and since the *velocity*, *v*, and the *height*, *h*, are the same at the beginning and at the end of the circulation, the difference in mechanical energy is represented by the difference in pressure *P*; (2) the *muscular pump*; as known from the anatomy, the veins are provided by *valves* that allow the blood to move in one direction only (i.e. towards the heart) when they are squeezed by muscular contraction; (3) the *negative* intrathoracic pressure (around the heart) due to the tendency of the lungs to collapse (we will study this in detail in the *Respiration* section of these notes).

As mentioned above 3000 ml of blood are contained in the venous circle. The volume of blood emitted by the heart at each systole is about 70–80 ml in the adult. If the *venomotor tone* would decrease of only 1% the return of blood to the heart would decrease by 30 ml, i.e. about half of the volume of the heart output pulse! In order to maintain the blood pressure *P* constant, in spite of the reduced blood input into the heart, the heart *frequency* must increase. We can then conclude in general terms that the blood pressure: $\Delta P = Resistance$ (PRU) $\times$ *Flow* is affected *directly* by a change of the *vasomotor tone* (i.e. by a change in cross section of the arterioles), and *indirectly* by a change of the *venomotor tone* through a change of blood input into the heart.

1.7.4 Pulmonary Circle

The circulation of blood in the lungs is affected by two factors: (*i*) the tension of lungs elastic structures, which tends to dilate the intrapulmonary vessels by pulling on their walls when the lung volume is increased; (*ii*) the pressure of air in the alveoli, which tends to collapse the vessels. The experiment illustrated in Fig. 1.41 shows how to investigate the combined effect of these two factors.

The question is: when the pulmonary volume is increased by increasing the pressure into the bronchus, the volume of blood within the lobe will increase due to the greater outward stretch on the vessel walls, or decrease due to the greater positive pressure of the air pushing on them? The result of these two opposite factors can be observed by the movement of the liquid into the two graduate horizontal tubes in Fig. 1.41 (eccentric: less volume in the lung vessel, concentric: more volume). The answer is: when the pressure of the alveolar air is increased and the lobe is inflated, blood is retrieved into the lung lobe. We can then conclude that the mechanical dis-

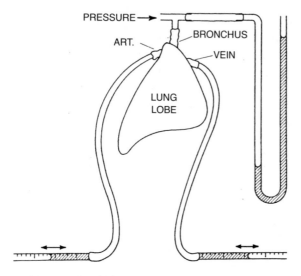

Fig. 1.41 Schema of the experimental set up used to determine the effect on the blood content in pulmonary vessels of (1) the alveolar pressure measured by a manometer connected to the bronchus and (2) the intravascular pressure, which can be changed by changing the height of the two horizontal tubes connected to the arterial and the venous ends of the blood vessels within the lung lobe (modified from Permutt S, Howell JB, Proctor DF and Riley RL *J. Appl. Physiol.* 16: 64, 1961)

Fig. 1.42 Schematic
representation of the forces
acting in an intrapulmonary
vessel surrounded by alveoli
(modified from Howell JB,
Permutt S, Proctor DF and
Riley RL *J. Appl. Physiol.*
16: 71, 1961)

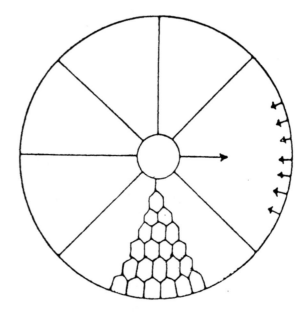

tension of the lungs causes a dilatation of the blood vessels (and of the bronchi, as we
will see) in spite of an increase of the air pressure around them. But why that should
be so? The central circle in Fig. 1.42 represents an artery or a vein surrounded by
alveoli connecting the vessel walls with the inside of the pleural surface or of another
limiting membrane. The air pressure ($P = force/surface$), in static conditions, is equal
everywhere, but will act on a smaller surface (s) at the middle (on the vessel walls)
and on a greater surface (S) on the outskirts (pleural surface). It follows that the force
f compressing the vessel ($f = P \times s$) is smaller than the force ($F = P \times S$) tending
to expand the lobe. However, in order to prevent rupture, the outward greater force F
must be counterbalanced by an equal and opposite inward force transmitted by the
alveoli to the vessel. It follows that the forces acting on the vessel are a smaller one f
caused by the alveolar air pressure, which tends to compress it, and a greater one F,
transmitted by the alveoli, which tends to expand it: the central vessel will then be
enlarged by the expansion of the lungs, in spite of the positive alveolar air pressure
tending to compress it, and liquid will move towards the lobe in the two horizontal
vessels of Fig. 1.41.

  From Fig. 1.42 we can deduce that the *capillary vessels* contained in the alveolar
structure transmitting the force from the pleura to the central vase will be flattened due
to the tension on their walls with the consequence that less blood will be contained
within them (about 100 ml in both lungs of an adult containing 1.7 l of blood in both
lungs, Fig. 1.26).

  In Fig. 1.43 the lung of a standing human is divided in three zones. The pressure
of the alveolar air, whose density is negligible ($d \sim 1$ g/l), is practically the same
in each of the three zones, whereas the pressure of blood (whose density is about

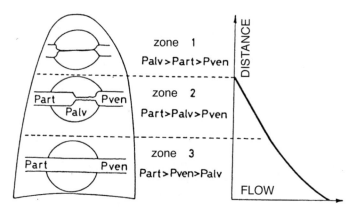

Fig. 1.43 Distribution of the blood flow within the lung *capillaries*. The lung is divided in three zones depending on the relative importance of the arterial, alveolar and venous pressures. The graph on the right shows the trend of the blood flow within the capillaries at the three different heights (modified by West JB, Dollery CT and Naimark A *J. Appl. Physiol.* 19: 713, 1964)

1000 times greater than that of air) increases from top to bottom (remember from the Statics that $dgh + P = $ constant, i.e. $\Delta P = -dg\Delta h$). At the top of the lung (zone 1 in Fig. 1.43) the capillaries in the alveoli are closed because blood pressure in them is less than the pressure of the alveolar air at both their arterial and venous extremities. As a consequence blood flow is nil in zone 1 (right end graph). The opposite is true in zone 3 where the capillaries are full open and blood flow increases to a maximum at the bottom of the lungs due to the greater hydrostatic pressure enlarging the capillary section.

A singular condition takes place at the middle of the lungs (zone 2) where the blood pressure is greater than the alveolar air pressure at the arterial extremity of the capillary and less at the venous extremity. At the point along the length of the capillary where the alveolar air pressure becomes greater than blood pressure, the capillary will collapse and the blood flow will stop; but, as soon as this happens, the pressure will rise to the level of the arterial pressure at its extremity, the capillary will reopen and the cycle will repeat continuously, leaving the venous segment of the capillary in a state of partial collapse. In this condition, called *flutter*, the flow of blood through the capillaries *depends on the difference between arterial pressure and alveolar pressure, independent of the pressure in the pulmonary veins*. After the resistance offered by the arterial segment of the capillary, the blood will *fall* in the venous reservoir and for this reason this condition is called *waterfall effect* (Fig. 1.44). Why is this so? The input flow into the capillary $\dot{V}_{in}$ must equal the output flow $\dot{V}_{out}$. The input flow is $\dot{V}_{in} = (P_{art} - P_{alv})/R_{art}$ where P_{art} is the arterial pressure, P_{alv} is the alveolar pressure and R_{art} is the resistance offered by the length of the capillary segment from its beginning to the point of partial collapse (arterial segment of the capillary). Note that for a given P_{art} and P_{alv} the resistance R_{art} is constant. The output flow is $\dot{V}_{out} = (P_{alv} - P_{ven})/R_{ven}$ where P_{alv} is the alveolar pressure, P_{ven}

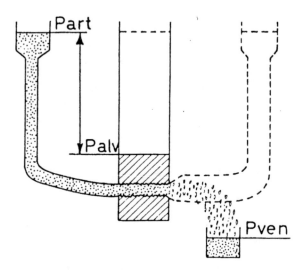

Fig. 1.44 Waterfall effect: the flow of blood through the lung capillaries depends from the difference between arterial pressure P_{art} and alveolar pressure P_{alv} (represented by the hydrostatic pressure of the central container). The venous pressure P_{ven} does not affect the flow, provided that it does not attain a value greater than P_{alv}. Flow stops when P_{alv} or P_{ven} attain values equal to P_{art} (interrupted lines) (from Riley RL *Ciba Foundation Symposium*. London: J.&A. Churchill Ltd., 1962)

is the venous pressure and R_{ven} is the resistance offered by the partially collapsed capillary segment (venous segment of the capillary). The length of the partially collapsed segment of the capillary is independent of P_{ven}, which cannot modify P_{art} nor R_{art} without opening completely the capillary. A change of P_{ven} will change R_{ven} by making more or less complete the closure of the segment of the capillary in a state of partial collapse, but will not affect $\dot{V}_{in}$. It follows that, since $\dot{V}_{in} = \dot{V}_{out}$, the venous resistance R_{ven} will not affect the blood flow in zone 2. Note, in addition, that the flow in zone 2 will increase linearly with the decrease in height h ("*Distance*" in Fig. 1.43) because $\Delta P = -dg\Delta h$.

1.7.5 Coronary Circle

The blood flow in the coronaries depends on two factors: (*i*) the resistance that the flow faces, due to the compression of coronary vessels by the contracting surrounding ventricular muscle, and (*ii*) the changes of blood pressure in the aorta from which they originate. Let's see how these two factors compensate each other (Fig. 1.45). The upper graph represents the sphygmic wave; note that the systole, during which blood is emitted into the aorta (bottom graph), lasts less than the diastole (0.2 vs. 0.7 s in the first heartbeat). The middle graph shows the flow of blood through the coronaries; it can be seen that after an initial peak the volume decreases to a very

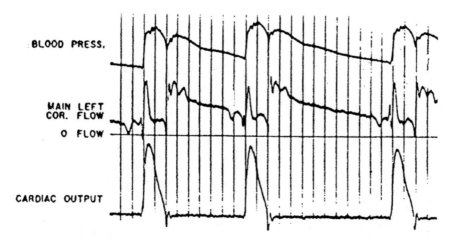

BLOOD PRESS.

MAIN LEFT
COR. FLOW

O FLOW

CARDIAC OUTPUT

Fig. 1.45 Upper graph, aortic pressure (sphygmic wave); middle graph, flow in the left coronary artery; bottom graph, blood flow in the aorta of a dog. Vertical time lines, 0.1 s (modified from Gregg DE, Khouri EM and Rayford CR *Circ. Res.*,16: 102, 1965)

low value during the systole to increase again during the long lasting diastole. The area below the middle graph ($\dot{V} \times t = V$) represents the *volume* of *blood irrorating the cardiac muscle, which is greater during the diastole than during the systole.* This is particularly important because the duration of the diastole *relative* to the total duration of cardiac cycle is greater the lower the frequency of the heart beats; it follows that *the lower the frequency the better is the oxygenation of the heart.*

1.7.6 Fetal Circulation

Fetus connection with the outside world takes place through the *placenta* where the oxygen uptake, the carbon dioxide emission and the metabolites exchanges occur. The lungs in the fetus are collapsed with the consequence that the blood vessels into them are constricted. A large resistance through the lungs and a small resistance towards the placenta characterize fetal circulation. Low oxygenated blood (58% of maximal oxygen saturation) flows into the placenta, instead of into the lungs, through the two umbilical arteries, and back from the placenta, where oxygenation takes place (80% saturation) through the umbilical vein into vena cava (*ductus venosus*) where it mixes with venous blood (left side of Fig. 1.46). This partially oxygenated blood flows into: (*i*) the right atrium and ventricle, (*ii*) the left atrium through the *foramen ovale*, and (*iii*) the pulmonary artery from which, instead of into the collapsed lungs, which offer a great resistance, blood flows into the aortic arch through the *ductus arteriosus*. The two ventricles of the heart that in the adult, breathing air, work *in series*, because the same volume of blood output by the right ventricle must be ejected by the left ventricle, in the fetal circulation the two ventricles work *in parallel*

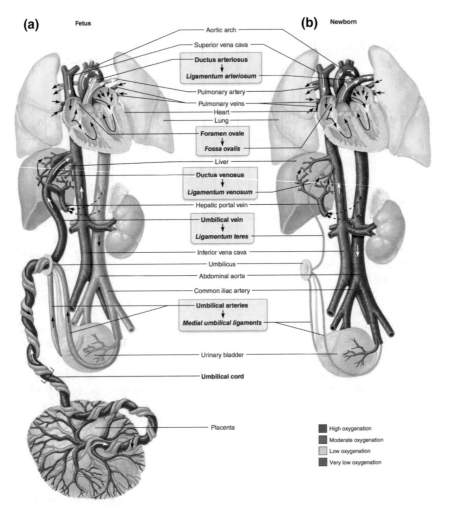

(a) Fetus

(b) Newborn

Aortic arch

Superior vena cava

Ductus arteriosus
↓
Ligamentum arteriosum

Pulmonary artery

Pulmonary veins

Heart

Lung

Foramen ovale
↓
Fossa ovalis

Liver

Ductus venosus
↓
Ligamentum venosum

Hepatic portal vein

Umbilical vein
↓
Ligamentum teres

Inferior vena cava

Umbilicus

Abdominal aorta

Common iliac artery

Umbilical arteries
↓
Medial umbilical ligaments

Urinary bladder

Umbilical cord

Placenta

■ High oxygenation
■ Moderate oxygenation
□ Low oxygenation
■ Very low oxygenation

Fig. 1.46 Schema of the fetal circulation before (left) and after birth (right) (from Marieb E and Hoehn K, *Human anatomy & physiology*, Pearson, Boston, 2013)

because both eject blood simultaneously into the aorta: the left one directly, the right one through the *ductus arteriosus*. In fact, in the fetus, the thickness of the walls of the two ventricles is equal, contrary to the adult where the thickness of the left ventricle wall is greater than that of the right ventricle due to the fact that the overall resistance and arterial pressure of the systemic circulation are greater than that of the pulmonary circulation.

What happens at birth (right side of Fig. 1.46)? The closure of the umbilical cord stops the input of oxygenated blood from the placenta: this results in an increase in the carbon dioxide in the blood of the fetus, which stimulates its respiratory center, leading to the first inspiration expanding the lungs: the liquid into the alveoli is substituted by air and the baby starts crying, i.e. breathing. The resistance in the vessels within the lungs falls, more blood reaches the left atrium through the pulmonary veins, the pressure in the left atrium increases and the foramen ovale begins to close thanks to a valve leaning on the foramen. The pressure in the pulmonary artery, which was high in the fetus because the lungs were collapsed, decreases. Before lungs expansion, the high pressure in the pulmonary artery was pushing blood into the aorta through the ductus arteriosus. After lung expansion, the flow in the ductus arteriosus is reversed and this contributes to its closure.

Now we will consider how the heart works, but since the heart is a muscle, we must consider the physiology of muscle first.

Chapter 2
Muscle, Locomotion and Heart

Abstract Definition of muscle positive work, isometric contraction and negative work. Location of containing and force transmitting structures. Neuromuscular transmission. Twitch, clonus and tetanus. Mechanical structures in series and in parallel. Muscle and sarcomere force-length and force-velocity relations with their mechanisms and functional consequences. Subdivisions and meaning of the heat produced by muscle. Locomotion: internal and external work. Walking. Running, hopping and trotting. Enhancement of muscular work induced by previous muscle stretching. The heart. Action potential. Pressure-volume changes in the cardiac cavities, in the aorta and in the pulmonary artery as a function of time during a heart cycle. The Starling law providing that the blood output of the left ventricle equals the blood output of the right ventricle. Pressure-volume diagram and work done by the heart ventricles. Cardiac output: stroke volume as a function of heart frequency. Efficiency of the heart: effect of stroke volume and heart frequency. Measurement of cardiac output: Fick's principle and dye method.

2.1 Introduction

What is a muscle? Muscle is a motor, i.e. a device capable to transform energy in another form of energy: specifically chemical energy into mechanical energy by doing mechanical work when it shortens (*concentric contraction = positive mechanical work = force × shortening*). Contrary to a petrol engine chemical energy is transformed into mechanical energy by muscle in an almost silent way (actually you may hear some sound when your head is lying on a cushion and you contract the masseter muscles). In addition whereas it is not possible to impede motion of a petrol motor (or electrical motor) without turning it off, muscle is capable to stay active without shortening (*isometric contraction*). Furthermore, as we will see, the muscle can remain active even when forcibly stretched (*eccentric contraction = negative mechanical work = force × lengthening*).

When muscle performs positive work the muscular force and its displacement have the same direction, when the muscle performs negative work force and displacement

© Springer Nature Switzerland AG 2019
G. Cavagna, *Fundamentals of Human Physiology*,
https://doi.org/10.1007/978-3-030-19404-8_2

have opposite directions: this happens when the force applied to the muscle is greater than the isometric force (*braking action of muscle*).

When muscles perform positive work and negative work in the every day activity? The heart performs solely *positive work* during contraction of atria and ventricles. In fish swimming and most kind of flying, mainly *positive work* is done by muscular force; this work is almost entirely dissipated as heat by friction against the water and the air. But, what about walking and running on the level at a constant step-average speed? The average gravitational potential energy and kinetic energy of the body remain, on average, constant and the work done against air friction and the soil (if no skidding takes place) is irrelevant. The answer is: most of the positive work done by the muscles returns in the muscles themselves while they are doing *negative work*. The *stretch-shorten* cycle of muscle, i.e. the succession of negative-positive work done by the muscular force, is a most common event in the everyday life. Also in a hummingbird, contrary to a heron, the frequency of wings beat is so high that probably negative work is done by muscular force to decelerate the wings at each beat.

2.2   The Location of the Motor

Figure 2.1 shows the different constituents of striated muscle. Without going in detail in the description of the anatomy of these structures, connecting the motor to the external load, we can pose the question: where actually is positioned the 'motor', i.e. the transformer of chemical energy into mechanical work? Muscle is made up by several muscular *fibers*, which in turn are an ensemble of *myofibrils* in parallel, with *sarcomeres* in series within them. The functional unit of muscle is the *half sarcomere*, which contains all the ingredients required for muscular contraction; these are represented by *actin and myosin filaments*, also called *thick and thin filaments*. The actin and myosin filaments are attached to the Z-disk and the M-line respectively and can slide over each other without changing their length. The structures between one Z-disk and the next represent a sarcomere. The thin filament is made up by *actin* molecules arranged as described in Fig. 2.1 together with the *troponin-tropomyosin* complex whose displacement relative to the actin molecules chain allows or inhibit attachment of the myosin molecule to actin, i.e. initiation or cessation of muscular contraction as it will be described below (Fig. 2.6).

The darker zone of the sarcomere, A band in Fig. 2.1, is made up by thick filaments of myosin that maintain the same length in the relaxed and contracting muscle. The lighter zone, *I* band in Fig. 2.1, is made up by thinner filaments of actin, which slide within the A band overlapping with the myosin thick filaments. The zone of non-overlap in the A band is lighter and is called *H* zone. The width of the zone of non- overlap in the lighter *I* band, decreases when muscle shortens because the thin filaments slide within the A band.

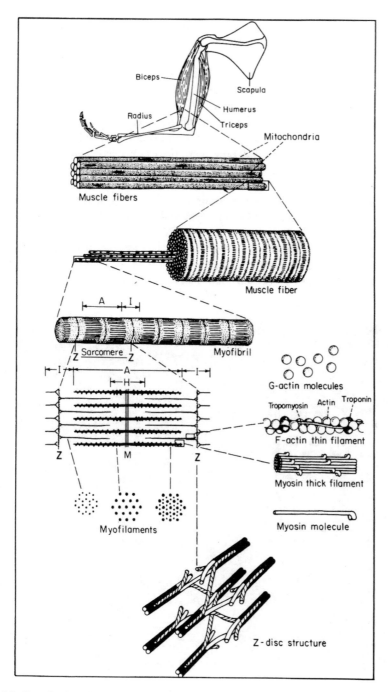

Fig. 2.1 Organization of striated muscle structure, showing the nomenclature applied to the bands and the proteins that make up the thick and thin filaments (from McMahon TA *Muscles, Reflexes and Locomotion*, Princeton University Press, 1984)

What causes the sliding of the thin filaments over the thick filaments when muscle shortens? Figure 2.2 shows isolated myosin molecules. In some of these is evident a long *tail* ending with two globular *heads*.

Figure 2.3 shows a schematic representation of the connection between one myosin head and the actin. Once a head of myosin comes in contact by thermal motion with a molecule of actin it attaches spontaneously to it attaining a high level of conformational potential energy. From this position the head tries to rotate spontaneously towards adjacent sites on actin forming a bound with a lower level of potential energy. If movement is prevented, the extensible elastic structure of the myosin tail will be stretched and force will be transmitted to muscle ends; in this case the head would oscillate back-forth between adjacent sites of actin. If rotation is allowed till the last possible site is attained, the myosin will remain attached to actin unless chemical energy (ATP) is used to detach it from actin. It follows that the "motor" in muscle is located in the point of contact between myosin head and actin molecule: chemical energy is used to detach otherwise spontaneously formed bounds. If the chemical energy required to perform this detachment is not available the muscle will remain contracted (*rigor mortis*: stiffness of death).

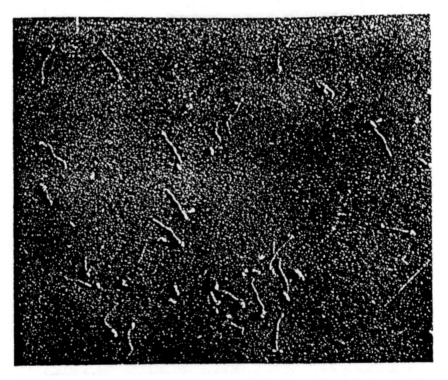

Fig. 2.2 Electron micrograph of isolated myosin molecules from rabbit skeletal muscle (from Elliott A, Offer G and Burridge K, *Proc. Roy. Soc. B* 193, 1976)

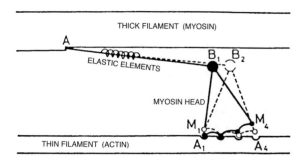

Fig. 2.3 Diagram of a cross-bridge emerging from the thick filament of myosin and attached to the thin filament of actin through elastic elements (A–B). The continuous line shows the head of myosin in a position with connections M_1–A_1 and M_2–A_2; the interrupted line shows a position attained by rotation of the myosin head by connections M_2–A_2 and M_3–A_3 (modified from Huxley AF and Simmons RM *Nature*, 233, 1971)

The mechanism of force and work generation described above for muscle is an example of a general condition found in nature: the transformation of Brownian motion into mechanical work at a constant temperature. The idea is to allow Brownian motion in one direction and hinder motion in the opposite direction by a stop. This is described in Feynman's ratchet-and-pawl (Vol. 1, Chap. 46, Lectures on Physics, 1963). The point is that the stop itself will be subjected to Brownian motion and an external source of energy is required to allow motion in one direction only. In case of muscle this external energy input is that of ATP.

2.3 Containing and Force Transmitting Structures

Muscle's contractile component is contained by *fasciae* around the whole muscle, *connective tissue* between muscular fibers, *sarcolemma*, the membrane of a singular fiber, containing *myofibrils* with filaments of actin and myosin (Fig. 2.1). At each end of the muscle fiber, the surface layer of the sarcolemma fuses with a tendon fiber, and the tendon fibers in turn collect into bundles to form the muscle tendons that then adhere onto bones. However also *within* the sarcomeres other passive structures exist having the goal to maintain the ordered arrangement of the myosin and actin filaments (Fig. 2.4). These structures consist of filaments of proteins of very high molecular weight, *titin* and *nebulin*, spanning the whole length of the sarcomere and *connecting the extremities of the myosin filaments with the Z line*. Their function is to stabilize the relative position of the myosin and actin filaments within the sarcomere. In fact, when the titin and nebulin filaments are broken by radiation, a misalignment between filaments takes place when muscle is activated (Fig. 2.4).

The titin and nebulin filaments correct a condition of inevitable instability within the sarcomere: in fact, it is practically impossible that the number of cross-bridges

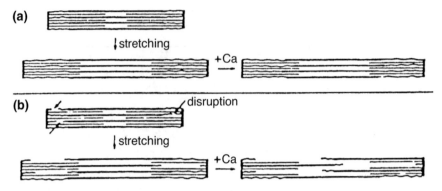

Fig. 2.4 Model of a sarcomere in which the ends of each myosin filament is connected to the nearest *Z* line by elastic filaments of giant proteins *titin* and *nebulin*. **a** Normal condition. **b** Specimen radiated in such a way to disrupt proteins of very high molecular weight (1,400,000–2,800,000) titin and nebulin. The disruption of these elastic elements due to radiation (arrows) causes disarrangement of the actin and myosin filaments when stretched or activated with calcium ions (from Horowits R, Kempner ES, Bisher ME and Podolsky RJ *Nature*, 323: 160, 1986)

between actin and myosin be *exactly* equal in the two half of the sarcomere. Sliding of actin-myosin filaments will then take place in the half sarcomere with accidentally a greater number of cross-bridges, but this result in a decrease of the number of cross-bridges on the half sarcomere of the other side, increasing the initial instability. This *positive feed-back* is corrected by the elastic strain of the titin and nebulin filaments, which prevent the initial drift and maintain the myosin filaments centered within the sarcomere.

All the elastic elements mentioned above, which are stretched by *lengthening* of the relaxed muscle, are *damped*, i.e. provided by poor elasticity and greater hysteresis (Fig. 1.20b); these are commonly called *parallel elastic elements*. In contrast, *undamped* elastic elements, improperly called *series elastic elements* because they are positioned in the tendons (but also in cross-bridges heads, and actin-thin filaments themselves), are stretched by the *force* exerted by muscle machinery. These undamped elastic elements have very small hysteresis, i.e. return most of the elastic energy stored during stretching. When talking about elasticity, e.g. in jumping, running etc. one refers to these undamped elastic elements.

2.4 Neuro-muscular Transmission

Muscle can be stimulated *directly* by means of electrical voltages on its surface bypassing the *neuro-muscular junction*, but physiologically muscle contraction results from an action potential of a motor neuron reaching the sarcolemma on the surface of the fiber. In short: when the action potential of the motor nerve reaches its terminal *motor endplate* on the muscle sarcolemma (neuromuscular junction), calcium ions (Ca^{++}) are allowed to enter the presynaptic membrane causing the release of the neurotransmitter *acetylcholine* (ACh) into the synaptic cleft. ACh attaches to the postsynaptic receptors on muscle's sarcolemma opening ions channels and allowing sodium ions (Na$^+$) to enter the sarcolemma cell (negative inside); this causes a depolarization of the membrane potential and eventually an action potential which runs along the myofibril causing muscular contraction. Na$^+$ concentration is then reduced by diffusion and by the enzyme *acetylcholinesterase*, the Ca^{++} ions are recaptured in the sarcoplasmic reticulum sponges and muscle relaxes. What follows explains how the action potential running on the sarcolemma on the *surface* of the fiber can elicit simultaneous contraction of muscle fibrils *inside* the fiber (Fig. 2.5).

The action potential causes a depolarization of the sarcolemma that, *at the level of the Z lines*, runs deeply into the fiber through the *transverse tubules*. The electrical

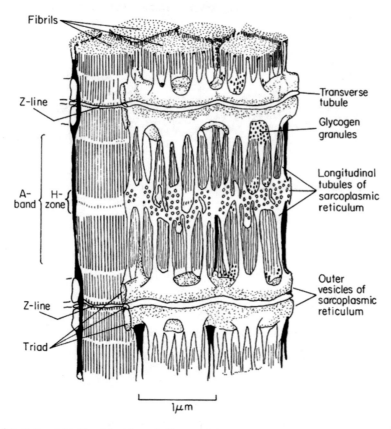

Fig. 2.5 In the right side of the Figure is illustrated the *sarcoplasmic reticulum* composed by the invagination of the sarcolemma at the level of the Z lines: the *transverse tubules*, which run parallel to two sarcoplasmic sacs in the adjacent sarcomere forming a *Triad*. The sacs continue into the *longitudinal* tubules, which run along the sarcomere (modified from Peachey LD *J. Cell Biol.*, 25, 1965)

depolarization of the transverse tubules affects the adjacent *outer vesicles and lon-gitudinal tubules of the sarcoplasmic reticulum*, which release calcium ions (Ca^{++}), causing muscular contraction. The mechanism by which calcium ions initiate and terminate muscular contraction is described in Fig. 2.6.

In between the double chain of actin molecules of the thin filament, a long molecule, the *tropomyosin*, prevents in the relaxed muscle the attachment of the myosin heads on the actin because it covers the attachment sites of myosin on actin. Three separate molecules, *troponin T, C and I*, forming a tight complex, are attached to one end of each molecule of tropomyosin. Attachment of calcium ions to troponin *C* causes the detachment of the inhibitory troponin *I* from seven actin molecules by rotation of the whole filament of tropomyosin through the action of troponin *T*. As soon as the 'active' site on the actin molecule is exposed, the myosin head sponta-neously attach to it reaching a maximum of conformational potential energy from which it tends to rotate spontaneously toward lower level of potential energy causing

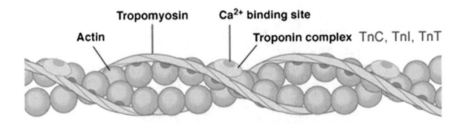

(a) Myosin binding sites blocked; muscle cannot contract

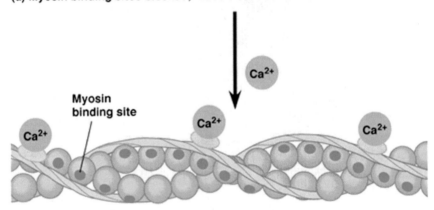

(b) Myosin binding sites exposed; muscle can contract

Fig. 2.6 Above: muscle is relaxed because, in absence of calcium ions (Ca^{++}) bound to the tro-ponin complex (troponin C, I and T), tropomyosin covers the attachment sites to myosin of seven actin molecules. Below: attachment of calcium ions to the troponin complex removes tropomyosin exposing the active actin sites (© Addison Wesley Longman, Inc., 1999)

muscle shortening. When sarcolemma repolarizes after the action potential, calcium ions are reabsorbed by the sarcoplasmic reticulum, tropomyosin returns to its resting condition covering the active sites on the actin molecules and the muscle relaxes.

2.5 Twitch, Clonus and Tetanus

When a muscle is stimulated *isometrically*, i.e. fixing its extremities at a constant distance, it develops a force that can be measured as a change in electrical resistance by means of *strain-gauges*, which allow to measure the force without an appreciable length change. The record, so obtained, is called a *myogram* (Fig. 2.7). In the whole muscle, the force developed can be regulated by applying stimuli of different voltages (both directly on its surface and indirectly through the motor nerve) to activate a different number of fibers. A stimulus high enough to stimulate all of the fibers of the whole muscle is called *supramaximal*. In the every day life muscle force is regulated by changing the number of active fibers. In what follows of these notes, whole muscle contraction will always be the maximal obtained using supramaximal stimulation. Contrary to the whole muscle preparation, the force developed by a *single* fiber is independent of the voltage of the stimulus; if the voltage is too low the single fiber will not contract: above a *threshold stimulus* the force developed is always the same independent of the stimulus voltage (*all or nothing law*).

When one action potential attains the neuron-muscular junction the isometric myogram measured is equal to that indicated by the interrupted line in Fig. 2.7 and is called a *twitch: characterized by a rapid increase followed by a slower decrease of the* force. If *maximal* repetitive stimuli are applied with an increasing frequency to the motor nerve (or directly to the muscle or single fiber surface), the force developed by the muscle (or by the single fiber) changes as described in Fig. 2.8. At low and intermediate frequency values, the force: (i) oscillates with the same frequency of the stimuli, and (ii) its average value is greater the greater the frequency. This condition

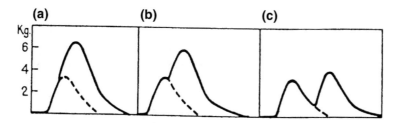

Fig. 2.7 Summation of responses in skeletal muscle. Isometric contraction curves (myograms) of mammalian nerve-muscle preparation. In each record the lower curve represents the response of the muscle to a single maximal stimulus to the motor nerve (twitch). The continuous thick line represents the response to the initial stimulus followed by a second stimulus. The second stimulus in *A* was applied during the rise in tension, in *B* at the height of contraction, and in *C* during relaxation (modified from Cooper S and Eccles JC, *J. Physiol.*, 1930, 69)

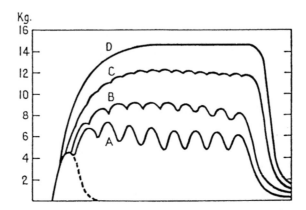

Fig. 2.8 Genesis of a *tetanus*. Response of mammalian force-muscle preparation to maximal stimulation through the motor nerve. Isometric contraction records. Lowest curve: response to a *single maximal* stimulation (single twitch). *A–C*: responses to *maximal* repetitive stimuli of increasing frequency. Curves *A–C* show partial tetani (*cloni*); curve *D* shows a full tetanus (modified from Cooper S and Eccles JC, *J. Physiol.*, 69, 1930)

is called *clonus* or *incomplete tetanus*. When the frequency of stimulation increases to a given value, the force attains a maximum maintained without oscillations; this condition is called *tetanus*.

The graphs in Fig. 2.8 are in apparent contrast with the *all or nothing law*: if above a *threshold stimulus* the force developed by each fiber is always the same regardless of the stimulus voltage, why increasing the *frequency* of the stimuli the force exerted by the fiber increases? The answer is: because the force developed by the contractile component is transmitted to the force transducer at the extreme of the fiber by undamped elastic elements represented by the cross-bridges themselves, the thin and thick filaments and the tendons. *When the force increases the undamped elastic elements lengthen and the contractile component shortens; when the force decreases the contractile component lengthen.* As we will see the force exerted by the contractile component *changes with the velocity of shortening and lengthening* (*force-velocity* relation, Fig. 2.18). Only when the force developed by the fiber is constant (condition *D* in Fig. 2.8) the contraction is really *isometric* within the fiber context and the force developed by the contractile component, the maximal exerted according to the *all or nothing law*, can be measured at the fiber extremities.

In agreement with the *all or nothing law* the force exerted by the fiber *contractile component* in isometric conditions attains always a maximum independent of the frequency of stimulation: the *active state*, ideally measured at the extremes of the contractile component *without interposition of elastic elements*. In Fig. 2.9 the continuous lines show: (i) the *active state*, assuming for simplicity that all the cross-bridges attach simultaneously to actin, and (ii) the corresponding force measured at the fiber extremities: the *twitch*.

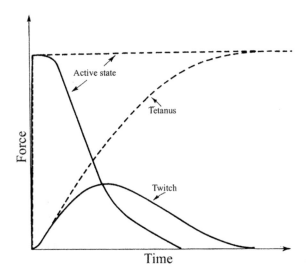

Fig. 2.9 The *continuous lines* indicate the force developed as a consequence of a *single stimulus* above threshold in isometric conditions: (i) measured at the extreme of the contractile component without interposition of elastic elements: *active state*, and (ii) measured at the extreme of the fiber, i.e. transmitted by elastic elements interposed between motor and force transducer: *twitch* (Figs. 2.7 and 2.8). The *interrupted lines* shows that if the stimuli frequency is increased to the limit required to attain a fusion of the individual contractions (as in Fig. 2.8 *D*) the force measured at the fiber (or muscle) extremities increases progressively up to the level of the *active state* (from Cavagna GA *Physiological aspects of legged terrestrial locomotion*, Springer International Publishing, 2017)

The relationship between the two continuous lines in Fig. 2.9 can be understood considering that, as it will be described subsequently, the contractile component exerts a force F lower than isometric when shortening and greater than isometric when lengthening (the *force-velocity* relation, Fig. 2.18). An increase in force, i.e. a positive slope dF/dt of the *twitch* curve in Fig. 2.9, implies that the undamped elastic elements lengthen and the contractile component, the "motor", shortens developing a force lower than isometric; vice versa a decrease in force. In the point where the two continuous lines cross in Fig. 2.9 the slope dF/dt is nil, i.e. the contractile component neither elongates nor shortens: the contraction is really isometric with the consequence that the force measured at the fiber extremities equals that of the active state.

When the frequency of stimulation is lower than that required to attain a continuous value of the active state, resulting in a fused tetanus as in Fig. 2.8 *D*, the result is a *clonus*, as that shown in Fig. 2.8 *A–C*. The genesis of a clonus is sketched in Fig. 2.10: the force measured at the extremes of the fiber increases when lower than active state and decreases when higher than it.

In conclusion, Figs. 2.9 and 2.10 show that the dependency of the force exerted by a fiber from the frequency of stimulation is not in contrast with the *all or nothing*

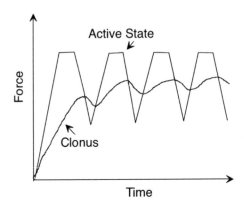

Fig. 2.10 Genesis of a *clonus*: a condition between that described by the full lines and the interrupted lines in Fig. 2.9. Four *active states*, following each a supramaximal stimulation of the fiber,
are separated by a time interval so that are not fused as in the interrupted line of Fig. 2.9. The force
exerted at the extreme of the fiber in this condition, *clonus*, increases when lower than the active
state and decreases when higher than it

law, but is merely due to the *site* where force is measured: indirectly through passive
elastic structures and not directly at the extremes of the motor.

From the experimental records described in Fig. 2.8 it appears that the only way to
have a continuous force (without oscillations) is to stimulate maximally the muscle
with a frequency of the stimuli high enough to attain a *fused tetanus*. However the
muscles of our body, in situ, can exert a continuous force, well below the maximal,
without oscillations. Figure 2.11 shows how this happens: the total resulting force
developed by the muscles in situ is lower than the maximal tetanic force and still
continuous (without oscillations) because several *cloni* of different *motor units (i.e.
the skeletal muscle fibers innervated by a single motor neuron) sum up out of phase* in
such a way that the resulting total force is continuous. In the experiment of Fig. 2.11
the cloni of only two motor units are summed up out of phase, actually in the muscles
in situ several motor units are involved: when the out of phase mechanism fails a
tremor appears. In conclusion, muscular force can be changed by increasing: (i) the
frequency of the action potentials in each motor unit, and (ii) the *number* of the motor
units recruited *out of phase* in order to have a continuous force-time relation. Note
that the out of phase contraction and relaxation of muscle fibers within the muscle
context is more suitable for the blood to flow than the constriction resulting from a
continuous contraction. Furthermore, note that the number of fibers innervated by a
motor unit is smaller the more precise is the movement required by each organ (e.g.
it is smaller in the finger than in the buttocks muscles).

All muscle physiology is based on two fundamental relationship: the *force-length*
diagram and *the force-velocity* diagram. Our goal is to explain these two relations on
the basis of the structural and physiological features described above.

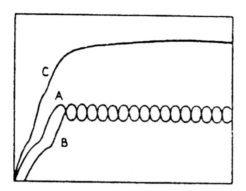

Fig. 2.11 Asynchronous discharge converting two cloni into a steady pull. Two *motor units* are stimulated at the same low frequency so that each unit responds with a clonus. If the interval between the onset of stimulation of *A* and *B* is suitably spaced the combined contraction of the units leads to a *steady* maintained contraction (curve *C*) (modified from Cooper S and Eccles JC, *J. Physiol.*, 69, 1930)

2.6 The Force-Length Relation

The findings that the functional unit of muscle is the *half sarcomere* consisting of actin and myosin filaments sliding relative to each other (Fig. 2.1) and that the number of force producing cross-bridges between them depends on filament overlap, poses the problem to investigate how the force developed by muscle depends on its length, i.e. on the actin-myosin overlap: this is the goal of the *force-length* diagram.

Figure 2.12 shows the experimental procedure used to determine the isometric force-length relation on the whole muscle, and Fig. 2.13 shows the corresponding graph obtained on an isolated toad sartorius muscle tetanically stimulated with supramaximal stimuli at different lengths. It can be seen that in this muscle (*sartorius*) the isometric tetanic force attains a maximum just at the length l_o where passive elastic elements begin to be set under tension. This is not always the case: in the *gastrocnemius* the passive elastic elements begin to be set under tension at a length smaller than l_o, in the *semitendinosus* at a length greater than l_o. In other words the contractile component is more 'protected' from excessive lengthening in gastrocnemius than in semitendinosus by the containing passive 'elastic' structures. Note also that the curve *P* in Fig. 2.13 was obtained during stretching the relaxed muscle; due to the large hysteresis of the passive elastic elements the *P* curve would be appreciably lower during shortening (Fig. 1.20b). The interrupted line *C* in Fig. 2.13 shows therefore the approximate trend of the contractile component alone at high muscle lengths.

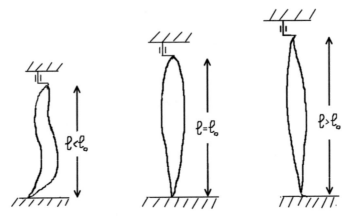

Fig. 2.12 Experimental set up to measure the *isometric* force-length diagram, i.e. the force exerted when the muscle is tetanically stimulated with supramaximal stimuli. The extremes of the muscle are connected at a fixed distance *l* and stimulated tetanically as in Fig. 2.8 *D*. The isometric force is measured by means of a force transducer fixed to an extreme of the muscle (a rigid bar with two attached *strain gauges*). In the Figure, three different lengths *l* are indicated below and above the resting length l_o, where the so called 'parallel' elastic elements containing the contractile component (fasciae, sarcolemma, etc.) begin to be set under tension. Lengthening the relaxed muscle beyond l_o, allows determining the force-length relation of these passive structures only

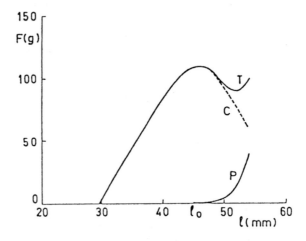

Fig. 2.13 Force-length diagram of a sartorius muscle during *static* isometric contractions: i.e. the force indicated by the continuous *T* line at each length was measured during a tetanic contraction at that length. The *P* continuous line indicates the force exerted by lengthening the relaxed muscle beyond l_o where passive elastic elements begin to be set under tension (toad sartorius 0.285 g, maximal length in situ 54 mm, minimal length in situ 46.8 mm). The interrupted *C* line is obtained by subtracting the *P* line from the *T* line (modified from Cavagna GA, Dusman B and Margaria R, *J. Appl. Physiol.* 24, 21:32, 1968)

2.6.1 Mechanical Structures in Series and in Parallel

Figure 2.14 shows *length* and *force* exerted by springs set in *series* and in *parallel*. The length change at the extreme of structures in *series* (i.e. attached one after the other as the sarcomeres within a fiber) is given by the *sum* of the length changes of each sarcomere, whereas the total force measured at the extreme of the series equals that exerted by *each* sarcomere. It follows that the length change in unit time of muscle, i.e. its *velocity* of shortening, will be greater the longer the muscle, i.e. the greater the number of the sarcomeres in series within its fibers. In contrast, the total length change of fibers in *parallel* within the muscle (i.e. one next to the other) will equal that of each fiber, whereas the total force exerted by all fibers in parallel will equal the sum of the forces exerted by each fiber (if the fibers are actually in parallel as in the sartorius muscle). It follows that the *power = force × velocity* will be greater the longer and the thicker is the muscle.

As mentioned above, the *functional unit* of muscle, i.e. the unit containing all the structures whose interaction is capable to explain the function of the muscle *in toto*, is the *half sarcomere*. One sarcomere is made up by two half sarcomeres in series. The cross-bridges within each half sarcomere are in parallel: the shortening of each half sarcomere equals that of each cross-bridge, whereas the force exerted by each half sarcomere is the sum of the forces of all cross-bridges and equals that of the half sarcomere in series with it. The functional unit, the half sarcomere, is the same in an athlete, in an untrained subject and in a frog.

Fig. 2.14 Left: three springs *in series*: the total force at the extreme of the series F_{tot} (arrow) equals that of each spring f, whereas the total lengthening ΔL_{tot} equals sum of the ΔL of each spring. Right: three springs in *parallel*: the total force F_{tot} (arrow) equals the sum of the forces of each spring f, whereas the total lengthening ΔL_{tot} equals that of each spring

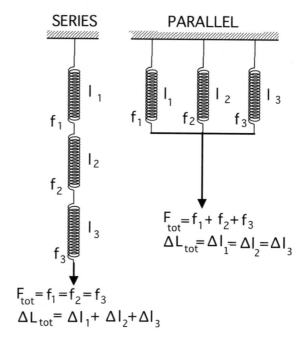

2.6.2   Sarcomere Force-Length Relation

What follows shows the liaison between anatomical and physiological observations. The values on the axis of the graphs reported until now depend on the size and the kind of the specimen used in the experiment (human, frog etc.). What follows shows how all the physiological observations reported above depend on a unique anatomic microscopic structure. In short, what follows shows the bound between a number of physiological observations and a unique anatomical mechanism. Whereas we should not memorize the values on the axis of Fig. 2.13 we should memorize those of Fig. 2.15.

The filaments of *myosin* containing cross-bridges are 1.6 μm long; in their center a segment 0.15–0.2 μm long does not contain myosin 'heads' but only 'tails' of the cross-bridges (Figs. 2.2 and 2.3). The thin filaments of *actin* in each half sarcomere are 1 μm long. It follows that when the sarcomere is $1 + 1 + 1.6 = 3.6$ μm long, the overlap between myosin and actin is nil and as a consequence the force developed upon stimulation is zero (1 in upper drawing in Fig. 2.15). If the actin filaments are allowed to slide over the thick filaments up to the point of maximum overlap between actin molecules and cross-bridges heads, the force developed upon stimulation increases, reaching a maximum at the sarcomere length of $1 + 1 + (0.15$–$0.2)$ μm ('tails' zone of myosin) $= 2.15$–2.2 μm (drawing 2 in Fig. 2.15). The finding that between 3.6 μm and 2.15–2.2 μm the force increases *linearly* shows that the cross-bridges are uniformly distributed over the myosin filament. If the sarcomere is allowed to shorten further until the two filaments of actin meet in the center of the sarcomere, the total sarcomere length will be 2 μm and the force will remain at its maximum because no further cross-bridges attach between actin and myosin (drawing 3 in Fig. 2.15). Shortening of the sarcomere below 2 μm causes first overlap between actin filaments (drawing 4) and subsequently compression of the thick filaments between the two Z lines (drawings 5 and 6), both of which will interfere with force development at the extreme of the sarcomere: the force falls to zero while inside structures are deformed. Actually a damage of the fiber was reported when contraction takes place at very short sarcomere lengths.

2.6.3   Functional Consequences of the Force-Length Relation

From the above description results that the muscle is capable to exert its maximal force over a short changes of its length, about 10% or little more. In other words, *the muscle is a greater producer of force, but not of displacement.* In the whole muscle the maximal force corresponds to a *stress (force/section)* ≈ 2–3 kg/cm^2. Figure 2.16 is aimed to emphasize how relevant is the force normally developed by muscles. To lift a 20 kg suitcase the upper arm muscles exert 140 kg and this high force is applied to the lower arm bone through a relatively thin tendon!

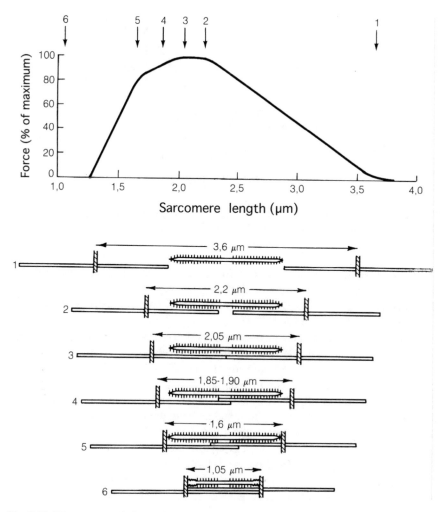

Fig. 2.15 The upper graph shows the isometric *force-length* relation of a single muscular fiber: on the ordinate the *force* is expressed as a percent of the maximal value attained, and on the abscissa the sarcomere length is given in µm. The lower part of the Figure shows, schematically, the overlap between myosin (thicker filament) and actin (thinner filament) at the sarcomere lengths indicated by the arrows in the upper *force-length* diagram (modified from Gordon AM, Huxley AF and Julian FJ, *J. Physiol.* (London) 184: 170–192 1966)

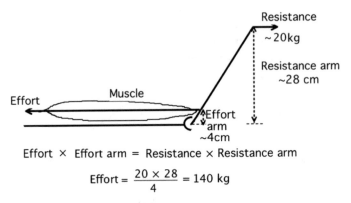

Effort × Effort arm = Resistance × Resistance arm

$$\text{Effort} = \frac{20 \times 28}{4} = 140 \text{ kg}$$

Fig. 2.16 Calculation of the force (Effort) developed by the upper arm muscle to lift a 20 kg weight (Resistance). The ratio Resistance arm/Effort arm = 7

In the movements of the everyday life we need not only force, but also displacement to perform work (force × displacement), e.g. during locomotion. Suitable displacements are provided by *class three levers* such as that illustrated in Fig. 2.16 at the expenses of a loss in force.

Another consequence of the force-length relation refers to the *stability* of adjacent sarcomeres in series. In the descending limb of this relation, i.e. from 2.2 to 3.6 μm sarcomere lengths (Fig. 2.15) a possibility of *instability* exists. Consider two adjacent sarcomeres in series: it is difficult to imagine that both have *exactly* the same length. The sarcomere even slightly longer than the other develops a smaller force because its overlap between actin and myosin, and as a consequence its number of cross-bridges, is less. As a consequence the shorter, stronger adjacent sarcomere, will lengthen the longer weaker one, decreasing further its actin-myosin overlap and its force: a condition of *positive feedback*. As described previously, a similar condition of instability also exists within each sarcomere where the position of the myosin filament is maintained at the center of each sarcomere by the *titin* and *nebulin* filaments (Fig. 2.4).

On the contrary, a condition of *stability* exists between sarcomeres in series in the ascending portion of the force-length relation. In fact a sarcomere shorter than its adjacent one will be weaker because of a greater constraint due to the overlap between actin and myosin filaments at its interior (4–6 in Fig. 2.15). It follows that the longer and stronger sarcomere in series would tend to lengthen the shorter weaker one, but in doing so will become weaker and the shorter one will become stronger opposing further lengthening: a condition of *negative feedback*.

The stability condition at lengths slightly shorter than the optimal one may play an important role in the hearth mechanics: if a ventricle receives an excessive volume of blood its muscular fibers will lengthen towards the optimal length and will be able to emit the excess blood volume during the subsequent systole (Starling law). However excessive lengthening, beyond the optimal length, into the descending limb, will cause a contraction weaker the greater the filling of the ventricle (heart failure).

Note however that other factors as the nervous system may interfere with the simple mechanism described above.

As it will be described below the instability between sarcomeres in the descending limb of the force-length relation is partially compensated by the other fundamental property of striated muscle: the force-velocity relation.

2.7 The Force-Velocity Relation

The relationship between applied load and velocity of length changes in isolated muscle specimens is determined experimentally using mechanical levers or, more recently, electronic feedback systems. Detailed description of these methods goes beyond the goal of these lesson notes. It is sufficient to understand that *isotonic contractions* (i.e. against a constant load) can now be attained with fair accuracy allowing to determine the relationship between velocity of the contracting muscle (or fiber) during shortening or lengthening against a *constant* imposed load: i.e. the *force-velocity* relation. If the force is constant (*isotonic*) the undamped elastic elements transmitting the force will maintain the same length, i.e. the length changes measured at the muscle extremities will equal those of the *contractile component only*.

Figure 2.17 shows the length changes of a single fiber tetanically stimulated, when the force at its extremities is suddenly reduced from the isometric value to a lower isotonic load. As expected the *first phase* observed is the recoil of the undamped elastic elements, simultaneous with the fall in force. More complex are the subsequent slower length changes. The second one, called *phase 2 isotonic shortening*, takes place against a constant force and is due to the completion of the 'rotation' of the myosin heads from their equilibrium position (Fig. 2.3). The *third phase*, corresponding to a slower shortening (sometimes even a slow lengthening), is probably due to the transition from cross-bridge distribution at the longer length to that at the shorter length. The *fourth phase* is the steady shortening against the new isotonic load and corresponds to cross-bridges attachment and detachment ('rowing'); the slope of this record corresponds to the *velocity* of the isotonic length change measured to construct the *force-velocity relation*. In the old times, when the apparatus used were slower, phases 1, 2 and 3 were not distinguished and described as a single phase, the recoil of elastic elements (phase 1); now the four phases are visible thanks to the fact the duration of the force step from isometric to isotonic is reduced to 100 μs.

The slope of the *fourth phase* is used to construct the force-velocity relation. Since the force changes with the length of the muscle (Fig. 2.15) the slope must be measured *at the same length for all loads*: the diagram force-velocity is an *iso-length relation* showing the effect of load on the velocity independent of the change in force taking place with length. This means that the slope of the length-time relation must be measured when the length change takes place across the *same* length of the specimen at all loads.

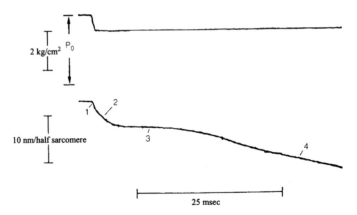

Fig. 2.17 The lower tracing shows the length changes of a single frog muscular fiber tetanically stimulated when the isometric force P_o is suddenly reduced from P_o to $0.8\,P_o$ (upper tracing). Four phases of shortening can be seen, whose significance is described in the text (modified from Huxley AF *J. Physiol.* 243, 1974)

If a constant load *greater* than isometric force is suddenly applied to the fiber, the trend of the graph of Fig. 2.17 is reversed: the length of the undamped elastic elements suddenly *increases* up to 1.5–1.8 the maximal isometric force (phase 1) followed after some transients by a steady lengthening (phase 4). The slope of the phase 4 record allows determining the *force-velocity of lengthening* relation.

2.7.1 Effect of Muscle Length

We have seen that if muscle is stimulated at a sarcomere length less than the optimal length l_o ~2 μm, the isometric force decreases because of the constraint due to the overlap of the filaments of actin at the center of the sarcomere (Fig. 2.15). The interrupted line in Fig. 2.18 shows that this internal impediment at $l < l_o$ also affects the velocity of shortening of muscle over the whole range of forces applied to it, from the maximal isometric down to zero.

If the muscle is stimulated at a sarcomere length greater than the optimal length, i.e. at $l > l_o$, the isometric force decreases because the number of attached cross-bridges is less due to the reduced overlap between actin and myosin (Fig. 2.15). The corresponding force-velocity relation is indicated by the dotted line in Fig. 2.18. It can be seen that, as expected, the velocity at $F \leq F_o$ is less at $l > l_o$ than at l_o. This is because the applied force is sustained by a smaller number of cross-bridges, with the consequence that the force per cross-bridge is greater and its velocity of 'rowing' is less. However as the applied force is reduced, the dotted line approaches the continuous line and both lines coincide at $F = 0$. To understand this trend, imagine a canoe in water with a different number of rowers. It is clear that the velocity of the

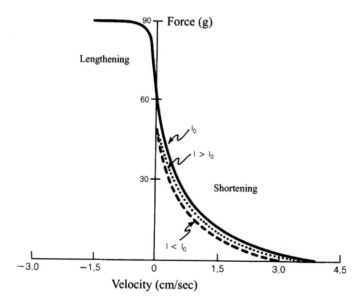

Fig. 2.18 Schematic representation of a force-velocity relation of a frog sartorius muscle at 0 °C. On the ordinate force is expressed in grams, on the abscissa velocity is expressed in centimeters per second. Positive values on the abscissa indicate the velocity of shortening, negative values the velocity of lengthening when stretching the contracting muscle. The continuous line refers to the resting length l_o, the broken line to a length less than l_o and the dotted line to a length greater than l_o. (Based on data of Aubert XM 1956, reported by Woledge RC, Curtin NA and Homsher E *Energetic aspects of muscular contraction.* Monograph of the Physiological Society, Academic Press, London, 1985)

canoe will be greater the greater the number of rowers, because the force per rower will be less. Now suppose to lift the canoe out of water: it is evident that the velocity of rowing will be independent of the number of rowers. In case of muscle this velocity, the maximal velocity attained by the dotted and continuous lines, corresponds to the maximal intrinsic velocity of ATPase detachment of cross-bridge from actin. In contrast at $l < l_o$, the maximal velocity of rowing will be less even when the external load is zero because the impediment to rowing is inside, not outside the canoe.

2.7.2   Mechanism of the Force-Velocity Relation

A. F. Huxley assumed that the cross-bridges emerging from the myosin filament M can attach to actin A within a given sliding distance x from the point of emergence from the thick filament and the point of attachment to the thin filament; he further assumed that the constant of velocity f of the myosin-actin reaction (the 'facility' with which attachment takes place) is nil when the distance x is nil (i.e. when the points of emergence and of attachment are one over the other), and increases linearly with filament sliding distance up to a point where the cross-bridge noes not 'reach' anymore actin and attachment can no longer exist. As in all chemical reaction, the possibility exists that the reaction is reversed: let's assume with a constant of velocity g, which changes with x as described in Fig. 2.19.

$$A + M \overset{f}{\underset{g}{\rightleftharpoons}} AM$$

The trend of the constants f and g are plotted in Fig. 2.19 as a function of the sliding distance x from the point of emergence of the cross-bridge and point of attachment to actin. It can be seen that in the range of x from zero to h the myosin bridges will tend to attach because $f > g$, whereas when the relative position of points of emergence and attachment x is reversed g attains a very large value, i.e. the bridges cannot attach spontaneously and if they are 'brought' into this zone they detach quickly. In the positive range of x the bridges favor shortening of muscle, whereas in negative range they oppose shortening.

During the isometric contraction bridges attach in the zero to h range only. When muscle is allowed to shorten two things happens: (i) less time will be allowed for bridge attachment to actin in the zero to h zone and (ii) some of the attached bridges are brought into a zone where they can only detach exerting, before detachment, a force that opposes shortening. Both these events will increase with the velocity of shortening, up to the maximal velocity of shortening is attained where the force falls

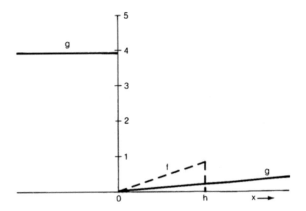

Fig. 2.19 Abscissa: distance x, along the major axis of the sarcomere, between point of emergence of the myosin bridge and point of attachment to the actin. Positive values of x refer to muscle shortening, negative values to muscle lengthening (stretched). The point where the elastic elements of the bridge are relaxed is indicated by $x = 0$, the maximal distance the bridge can reach is h. Positive values of x result in a force which tends to shorten the sarcomere, negative values of x result in a force which opposes sarcomere shortening. On the ordinate are given the relative values of the velocity constants for bridge binding, f (interrupted line) and detachment, g (continuous line) (from Huxley AF *Prog. Biophys. biophys. Chem.* 7, Pergamon Press, The Macmillan Co., New York, 1957)

to zero because the cross-bridge distribution exerting a positive force equals that exerting a negative force (as indicated in Fig. 2.20).

As mentioned above, a condition of instability exists between sarcomeres in series in the *descending limb* of the *force-length relation*: a shorter sarcomere, stronger due to a greater overlap between filaments, occasionally adjacent to a longer sarcomere weaker due to a lower overlap, will further lengthen the adjacent one increasing the initial disequilibrium (positive feedback). The force-velocity relation tends to correct this condition of instability because the force exerted during the velocity of lengthening by the weaker sarcomere is greater than the force developed during

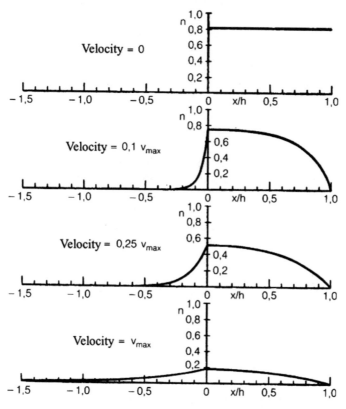

Fig. 2.20 Abscissa: the distance, x, along the major axis of the sarcomere, between point of emergence of the myosin bridge and point of attachment to actin is expressed as a fraction of h, the maximal distance a bridge can attain for binding to actin. Ordinate: the proportion n of the sites on the actin occupied by a bridge: for example, $n = 0.8$ means that 80% of sites at disposal are occupied. Note that the proportion depends on the distance x, as indicated in each graph, and from the velocity of shortening, as indicated from top to bottom by the different graphs: with increasing velocity of shortening, an increasing fraction of bridges are drawn in the opposite side of their point of emergence, with the consequence that they tend to impede sarcomere shortening and limit the maximal velocity of shortening (from Huxley AF *Prog. Biophys. biophys. Chem.* 7, Pergamon Press, The Macmillan Co., New York, 1957)

the velocity of shortening by the stronger sarcomere (Fig. 2.18). However when the difference in velocity between adjacent sarcomeres is very small, as in the isometric contraction, the stronger sarcomeres slowly lengthen the weaker ones causing a slow increase of the isometric tetanus at high sarcomere length (*creep*).

2.7.3 Power

Let's consider first the *force-velocity of shortening relation*. The meaning of this relation is expression of a *general property of all motors*. For example: a car will go slowly uphill developing a great force and faster on the level developing a low force. The product of the force time the velocity of shortening is the *power* developed by the motor. If the power were constant the product *Power = Force × velocity of shortening* would represent a hyperbola with power attaining infinity when the velocity is zero and vice versa. Actually the relationship is that of a hyperbola with *translation of axis* (equation of A. V. Hill):

$$(F + a)(v + b) = (F_o + a)b = \text{constant}$$

showing that when the velocity v is zero the force attains its maximum *isometric* value F_o and vice versa when the force is zero the velocity attains its maximum value, v_{max}. The maximal power is attained at 1/3 v_{max} corresponding to 1/3 F_o, and amounts to 0.1 $F_o v_{max}$. The maximal efficiency of muscle (positive mechanical work/chemical energy consumption) is attained at 1/5 of v_{max} corresponding to 1/2 F_o and amounts to 0.20–0.25. The choice between these two possibilities, maximal power or maximal efficiency, depends on the kind of exercise to perform. In a short race at the maximal speed 1/3 F_o should be preferred. In a long lasting journey with no food at disposal the 1/2 F_o should be preferred. Bike gearshifts allow muscles to contract near the optimal velocity ($1/3v_{max}$–$1/5v_{max}$) independent of the slope (Fig. 2.21).

2.7.4 Positive Versus Negative Work

The continuous line in Fig. 2.18 shows that (i) the force exerted by a maximally stimulated muscle during shortening (positive work) is lower than that developed during lengthening (negative work) and (ii) that the difference depends highly on the velocity of shortening and lengthening. The greater the velocity the greater the difference in force. For example: going upstairs requires a greater activation of muscles then going downstairs at the same velocity because each fiber, when stretched, exerts a much greater force. However suppose to go upstairs and downstairs at the maximal possible speed, but with a large load on your shoulders: the velocity of muscle contraction will be small in both cases and the force during positive work would almost

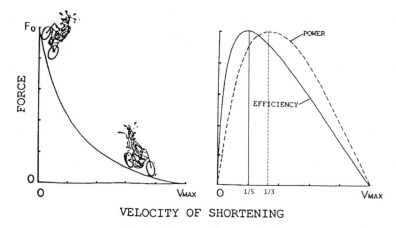

VELOCITY OF SHORTENING

Fig. 2.21 Left: No bike gearshifts! Right: Power and efficiency changes with velocity of muscle contraction: maximal efficiency is attained at 1/5 of the maximal velocity of shortening; maximal power is attained at 1/3 of the maximal velocity of shortening

equal that during negative work: i.e. the amount of muscle activation would be about equal in both cases. In other words: at a given load, the ratio between the number of active muscular fibers and relaxed muscular fibers is greater the greater the velocity of stretching and shortening.

2.7.5 Enhancement of Muscular Work Induced by Previous Muscle Stretching

Suppose trying to reach a fruit on a high branch of a tree, what would you spontaneously do? Starting from a squatting position and then jump or starting from the upright position, flexing your legs to the squatting position and then jump? Obviously the second case. This is because the *positive work done immediately after stretching the contracting muscle is greater than the positive work done without previous stretching*. Let's see why.

As described above, the *force-length* diagram (Fig. 2.13) and the *force-velocity* diagram (Fig. 2.18) are two landmarks of muscle physiology: the *dynamic force length diagram* is a combination of the two.

The independent variable in the isometric force-length diagram is the length, in the isotonic force-velocity diagram the independent variable is the force, in the dynamic force-length diagram the independent variable is the velocity: one end of the muscle or of the fiber, tetanically stimulated, is attached to a motor moving towards or away from it at a given velocity independently of the force exerted by the specimen.

We can describe three types of dynamic force length diagrams obtained during: (i) shortening only, (ii) stretching only and (iii) stretching followed by shortening.

In Fig. 2.22 it can be seen that when the isometric contracting specimen is allowed to shorten by the motor at a given imposed speed, it exerts during shortening a force lower than the isometric force; the force during shortening is greater the lower the speed (in 2 is greater than in 3, see 'shortening' in Fig. 2.18). When the speed of shortening is very high (a 'step' release as 1 in Fig. 2.17) the only structures that recoil are the undamped elastic elements set under tension during the isometric contraction (because the time allowed for cross-bridge attachment and detachment is too short): in this case the dynamic force-shortening diagram approaches the force-length relation of the undamped elastic elements only.

When the isometric contracting muscle is stretched while active the force measured during stretching increases (somewhat independent of speed of stretching, see 'lengthening' in Fig. 2.18) well above the isometric value (4 in Fig. 2.22) and when the contracting muscle is maintained active at the stretched length, it reverts more slowly towards the isometric value (*stress-relaxation*: 5 in Fig. 2.22).

If the contracting muscle is allowed to shorten immediately after stretching, the area below the shortening curve, i.e. the positive work done, is greater than when it is allowed to shorten the same distance, from the same length and at the same velocity from a state of isometric contraction. In other words, the succession of negative-positive work, which often occurs in Nature, results in an *enhancement of the positive work done immediately after stretching the contracting muscle* (Fig. 2.23). The simplest explanation of this finding is the following: since the force attained at the end of stretching is greater than the isometric force at the same length, more elastic energy is stored within the undamped *elastic* elements and this additional energy is released during subsequent shortening making the work done after stretching (from a greater force) greater than that done from a state of isometric contraction (lower force). This however is not the only explanation for the greater amount of work done after stretching: the contractile component itself is enhanced by previous stretching (Fig. 2.24).

Fig. 2.22 Dynamic force-length diagrams superposed on the isometric force-length diagram (Fig. 2.13). 1—Isometric contraction; 2–3 Shortening from a state of isometric contraction at a constant velocity, lower in 2 than in 3; 4—Stretching the contracting muscle; 5—Stress-relaxation taking place when the muscle is maintained active at the stretched length

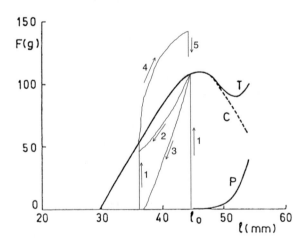

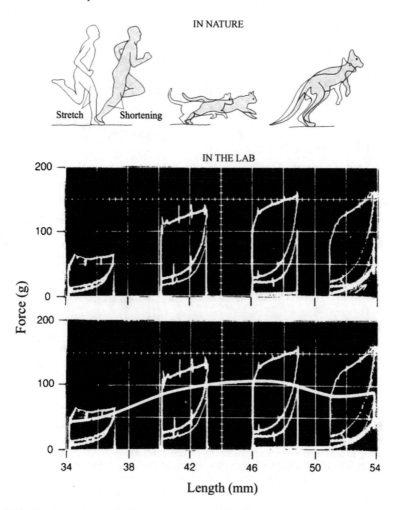

Fig. 2.23 The dynamic stretch-shortening cycle, which often occurs during locomotion (upper panel), is reproduced experimentally on the isolated muscle (toad sartorius) at four different lengths and compared to dynamic force-length diagrams measured during shortening from a state of isometric contraction (lower panel). In one case the muscle was tetanized isometrically at the shorter length, forcibly stretched while active to the greater length and allowed to shorten immediately after stretching. In the second case the relaxed muscle was lengthened to the same length attained during stretching, tetanized isometrically at this length and then allowed to shorten the same distance at the same velocity. Each couple of the dynamic force length diagrams determined at four different muscle lengths (abscissa) are shown in the bottom part of the Figure superposed to the static force-length diagram (Fig. 2.13) obtained by connecting the isometric force developed before stretching and shortening. The areas underlying the shortening curves represent the positive work done: this is greater when shortening begins immediately after stretching. Spikes are artifacts due to stimulation (modified from Cavagna GA, Dusman B and Margaria R *J. Appl. Physiol.* 24: 21, 1968)

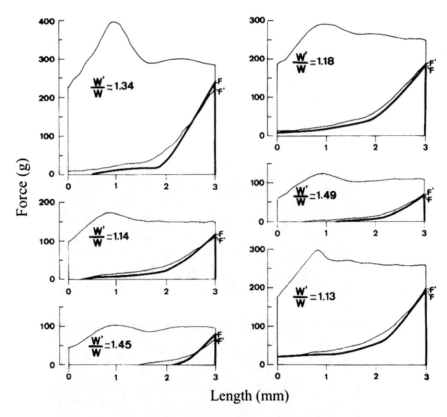

Fig. 2.24 Dynamics *force-length* diagrams on toad sartorii. The *heavy line* corresponds to an active shortening of 3 mm from the isometric force *F*. The *upper thin line*, starting from zero corresponds to stretching of the tetanized muscle. At the end of stretching, about 20 ms were allowed to elapse and the force fell to *F'*. Thereafter the muscle was allowed to shorten 3 mm (*lower thin line*). Since *F'* is less or equal to *F*, the excess work *W'* − *W* must be due to the contractile component only (modified from Cavagna GA, Dusman B and Margaria R *J. Appl. Physiol.* 24: 21, 1968)

   The enhancement of the contractile component induced by previous stretching of the contracting muscle is also shown by the finding that it can shorten against a *constant* force equal to the maximal isometric force at the same length possibly thanks to a greater conformational energy attained by cross-bridges (e.g. Cavagna GA, Mazzanti M, Heglund NC and Citterio G, *Am. J. Physiol.* 251: C571, 1986) and by the attachment of the second head of each bridge induced by stretching (Brunello E, Reconditi M, Ravikrishnan E, Linari M, Yin-Biao S, Theyencheri N, Panine P, Piazzesi G, Irving M and Lombardi V *Proc. Natl. Acad. Sci.* USA 104: 20114, 2007)

2.8 Heat Produced by Muscle

2.8.1 Meaning of the Heat Produced by Muscle

Let's consider a chemical reaction taking place in a test tube without particular arrangements capable to turn part of the chemical energy into work. In these conditions, the variation of *internal energy* of the system should be found totally as heat. The variation of the internal energy is defined in thermodynamics as $\Delta E = $ energy of products (E_2) − energy of reactants (E_1). This is a negative number in an exothermic reaction because the system is depleted of energy. The heat produced is on the contrary expressed as a positive quantity in calorimetry: *heat produced* $= -\Delta E$. Since the occurrence of double negative signs is font of confusion, one prefers to adopt, in the thermodynamics applied to the muscle, the symbol $\bar{\Delta}$ instead of $-\Delta$. Therefore $\bar{\Delta} = -\Delta$ and, in conditions of maximal inefficiency (in the test tube): *heat produced* $= \bar{\Delta}E$.

However, if during the reaction in the test tube some gas is produced and this gas expands, work $\int P\Delta V$ will be done by the system against the atmospheric pressure P. Work will be done on or by the atmospheric pressure even when, without gas production, a change in volume ΔV of the system takes place. Even in this case work is done by the system due to an increase of its volume, a fraction of the energy will be transformed into work and not into heat, and as a consequence the heat production will be less: *heat produced* $= \bar{\Delta}E - \int P\Delta V$.

The heat produced (or absorbed) during a chemical reaction (at constant temperature and pressure) when no possibility is given to the chemical reactants to perform work, except the work bound to a change in volume, is called *enthalpy*: ΔH. Since the change in volume of muscle during contraction and relaxation is negligible, it is not necessary to maintain the distinction between internal energy and enthalpy, i.e.:

$$\bar{\Delta}H \approx \bar{\Delta}E = \textit{heat exchange in conditions of maximal inefficiency of the system}$$

In presence of arrangements capable to capture part of the internal energy (or enthalpy) of the chemical reactants and to turn it into mechanical work, the enthalpy will not be totally turned into heat. A fraction of the energy will be found as work and the heat production assuming an exothermic reaction will be lower:

$$\textit{heat produced} = \bar{\Delta}H - \textit{work done } (W)$$

If the reaction were endothermic ($\bar{\Delta}H < 0$) it would absorb even more heat in case it would be allowed to perform work.

In principle one may imagine to transform completely the enthalpy of the chemical reactants into work. Is it possible to turn all of it into work? Is it possible to set equal to zero the heat produced and write:

$$\bar{\Delta}H = \textit{work done?}$$

The answer is: usually not. Even assuming that the motor works with 100% efficiency, a quantity of heat that could be (always according to our signs convention) positive, i.e. produced by the system, or negative, i.e. absorbed by the system, usually accompanies the process. This heat exchange that accompanies *reversibly (if it is released from A to B, is absorbed from B to A: i.e. no dissipation into heat takes place)* an ideal process taking place through equilibrium states is intimately bound to the structural changes of the system that accompany the transformation. It is given by the product of the absolute temperature T during the transformation times the changes in *entropy* taking place within the system as a consequence of the transformation itself: ΔS = entropy of products − entropy of reactants. We will use $\bar{\Delta}S$ = entropy of reactants (S_1) − entropy of products (S_2). A positive value of $T\bar{\Delta}S$ corresponds to a *reversible production* of heat (the reaction is exothermic, muscle warms up and heat passes from the system, muscle, to the surroundings). It follows that if the entropy of the reactants is greater than that of products $(\bar{\Delta}S > 0)$ a fraction of the enthalpy of the reactants will not be free to translate into work but will be found as heat produced reversibly: $T\bar{\Delta}S$. In this case the fraction of enthalpy, or of internal energy, free to translate in reversible conditions into work is only a fraction of the total enthalpy of the reactants: to this fraction in fact is given the name of *free energy, F.* The maximum work done in "*reversible*" conditions will then be:

maxmimum work $= \bar{\Delta}F = $ *free energy of reactants − free energy of products*

A reaction will take place spontaneously only if the free energy of reactants is greater than that of products, i.e. if $\bar{\Delta}F > 0$. In *reversible conditions*, therefore, the heat exchange will be positive (produced), nil or negative (absorbed) depending if the entropy of the reactants is greater, equal or less than that of the products.

In case that $S_1 > S_2$ $(\bar{\Delta}S > 0)$, the production of heat in reversible condition will be

$$heat\ produced\ (reversibly) = \bar{\Delta}H - \bar{\Delta}F = T\bar{\Delta}S$$

If the entropy of products equals that of reactants $(\bar{\Delta}S = 0)$, all of the free energy of reactants will turn in reversible conditions into work without heat production:

$$heat\ produced\ (reversibly) = \bar{\Delta}H - \bar{\Delta}F = T\bar{\Delta}S = 0.$$

If, on the contrary, $S_1 < S_2$ $(\bar{\Delta}S < 0)$, more than 100% of the enthalpy, or internal energy, of the reactants will be able, in reversible conditions, to turn into work: as if some energy would set free (because of structural modifications of the system) adding to that initially present in the reactants to be reversibly transformed into work. In this case one would observe absorption of heat by the system:

$$heat\ produced\ (reversibly) = \bar{\Delta}H - \bar{\Delta}F = T\bar{\Delta}S < 0.$$

In case the process would not take place in reversible conditions, but, as usually happens, the free energy would turn into work with an efficiency less than unity then,

$$heat\ produced = T\bar{\Delta}S + (\bar{\Delta}F - W)$$

The heat produced, which is measured during muscle activity, consists therefore of two parts:

1. $(T\bar{\Delta}S)$: a fixed quantity that can be large or small, positive, negative or nil, which appears independently of the way in which the reaction takes place and that is *reversible* with the reaction.
2. $(\bar{\Delta}F - W)$: the fraction of free energy that is degraded *irreversibly* into heat; which is always positive (or zero, in reversible conditions). The entity of the degradation into heat depends on the efficiency of the mechanism coupling the two energy forms, i.e. on the motor; it depends, in addition, on the velocity with which the process takes place.

All measurements of heat production must take into account these two parts. If the efficiency of the transformation is 100%, then $W = \bar{\Delta}F$, and the only heat exchange is that due to the entropy change.

If at the end of the process the state of disorder of the system equals the initial state of disorder, $\bar{\Delta}S = 0$:

$$\bar{\Delta}F = heat\ produced + W = \bar{\Delta}H$$

This occurs in muscle *at the end of a complete time interval including contraction, relaxation and recovery in oxygen*: in these conditions only it is possible to measure the efficiency of mechanical work production as the ratio:

$$efficiency = (work\ done)/(free\ energy\ used)$$
$$= W/(heat\ produced + W)$$
$$= W/\bar{\Delta}F$$

As the physical meaning of a lack of 'elasticity' is the 'hysteresis' (Fig. 1.20), the physical meaning of 'fatigue' is an 'entropy increase', i.e. the system is not returned to its initial condition (e.g. lactic acid production after a race etc.).

2.8.2 Subdivisions of the Heat Produced by Muscle

These are schematically represented below:

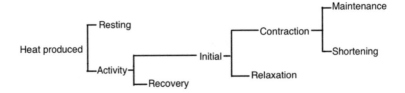

The *resting heat* of relaxed muscle is negligible and is required to maintain its chemistry unaltered with time. When muscle is stimulated the *activity heat* is produced in a short range of time, *initial heat* (of the order of seconds) and more slowly, the *recovery heat* (of the order of minutes) released when the system returns to its initial conditions ($T\bar{\Delta}S = 0$), i.e. when ADP is transformed back into ATP using oxygen. The *initial heat*, in turn, is produced both during muscular *contraction* and *relaxation*. The heat produced during muscle contraction is lower when the muscle is kept in isometric condition, *maintenance heat*, and is increased when the muscle is allowed to shorten, *shortening heat*.

When a single action potential reaches a muscular fiber the *activation heat A* is produced; the sum of more activations heat results in the *maintenance heat*, $M \cong \Sigma A$. During shortening another, smaller term is added: the *shortening heat*, $a\,x$ (where a has the dimension of a force and x is the shortening), i.e. *total heat* $= M + a\,x$. It is

Fig. 2.25 Schema of *initial heat* production during an isometric and isotonic contraction. Two conditions are illustrated during relaxation after an isotonic contraction: 1. When the muscle is kept at the final length reached during shortening, 2. When the muscle returns to its initial length and the mechanical energy output during shortening evolves as *relaxation heat* within the muscle (from Keele AC and Neil E *Samson Wright's Applied Physiology*, London, Oxford University Press, 1961)

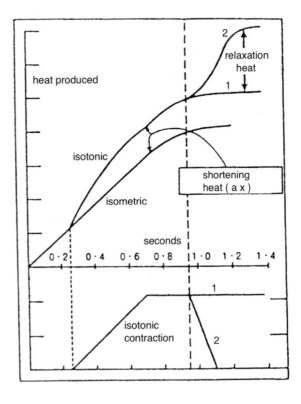

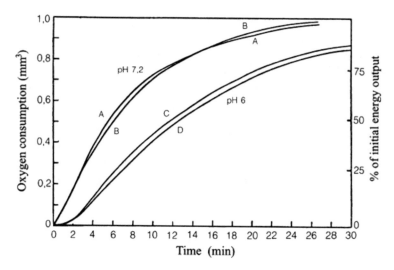

Fig. 2.26 Comparison between *recovery* heat production expressed as a % of the initial energy (right hand ordinate) and oxygen consumption in aerobic conditions (in mm³, left hand ordinate, as a function of time (in minutes: abscissa). The curves obtained at *pH* = 7.2 refer to normal condition: A: oxygen; B: heat. The curves obtained at *pH* = 6 refer to condition where metabolism is affected by an excessive acidity: C, heat; D, oxygen (from Hill DK *J. Physiol.* 98, 1940)

important to realize that M to maintain muscle contracted is much greater than ax added when muscle is allowed to shorten.

The energy set free when muscle shorten doing work, for example lifting a weight, is called *initial energy*: $E = M + ax + Fx$, where F is the force exerted during shortening. The term $(F + a)x$ is called Fenn effect showing that the energy the muscle sets free is not only that ready during the isometric contraction, but is *increased* when the muscle is allowed to shorten. Heat production during *relaxation* is: (i) *zero* if the work done Fx during shortening is stored *outside the muscle* as potential energy (e.g. gravitational potential energy, if a weight lifted by muscle from the ground is left on a table), or (ii) Fx, if the work done by the muscle is reabsorbed by muscle during relaxation (e.g. if the weight is lowered by muscle from the table to the ground (Fig. 2.25)). Even in the first case however some heat is produced during relaxation due to the energy set free by the undamped relaxing elastic elements previously set under tension during muscular contraction.

Let's now consider another case: *heat produced during forcible stretching of the contracting muscle*. Since in this case mechanical energy *enters* the muscle, one would expect, on purely physical terms, that the heat production would equal the maintenance heat *plus* the *negative* work done by the muscular force. Therefore it amazes that in some cases the heat production is *equal or even less* than during an isometric contraction. This finding can be explained by assuming that during forcible stretching part of the work done on the muscle is stored by the contractile machinery because the cross-bridges are brought by stretching to a higher level of chemical (conformational) energy.

From what described above it is clear that after the initial heat the system is not returned to its initial conditions. This is done more slowly when the *recovery heat* is produced and oxygen is absorbed by muscle; actually the recovery heat is the thermal accompaniment of the oxygen consumption required to return the system to its initial conditions (i.e. ADP $\to$ ATP). It turns out that *the recovery heat equals the initial energy output* but takes minutes instead of seconds to complete (Fig. 2.26). The *efficiency of muscle* $= Work/M + ax + Fx + recovery\ heat = Work/2(M + ax + Fx) = 0.20$–$0.25$. The so-called *initial efficiency* $= Work/M + ax + Fx \cong 0.5$ is not correct in physical terms because the system is not returned to its initial conditions, i.e. possibly $\overline{\Delta}S \neq 0$.

2.9 Locomotion

Figure 2.27 shows the weight specific energy expenditure per unit distance (ordinate) as a function of body weight (abscissa) during locomotion. It can be seen that energy expenditure decreases with body mass and that *for a given body mass is greater in terrestrial locomotion, intermediate in flight and minimal in swimming*. This is an apparently astonishing finding because in terrestrial locomotion the body is supported by the ground, the work done against air friction is small and the work done against ground friction is nil (except skidding takes place) whereas in flight the body must be actively maintained airborne by muscular contraction and in swimming the friction against the water is much higher than that against the air (the density of water is 1000 times greater than that of air).

Locomotion results from the interaction of a *motor*, the muscular system, and a *machine*, the lever system of the limbs. The minimum, inevitable *work* to maintain motion of an object through a medium is given by the product of the *displacement* of the object times the *drag*, the frictional resistance of the medium through which motion takes place. As illustrated in the schema of Fig. 2.28, the *overall efficiency* of locomotion is given by the ratio between this work and the chemical energy expenditure to sustain muscular contraction.

The overall efficiency in turn depends on *muscular efficiency* and the efficiency with which the lever system is capable to transform the positive work furnished by the muscles into displacement of the object through the medium in which the object is immersed. We call this *propulsive efficiency*, which is greater in a wheel, in flying and in swimming than in legged terrestrial locomotion. On the other hand the muscular efficiency is equal in human as in a frog (0.20–0.25). Why propulsive efficiency is so small in legged terrestrial locomotion?

Fig. 2.27 The energy expenditure to maintain locomotion is plotted as a function of body weight. The three curves refer to swimming of fishes, flight of birds and insects, and terrestrial locomotion of mammals and a lizard (from Tucker VA, In Bolis L (eds) *Comparative physiology*, North-Holland Publishing Company, Amsterdam London, 1973)

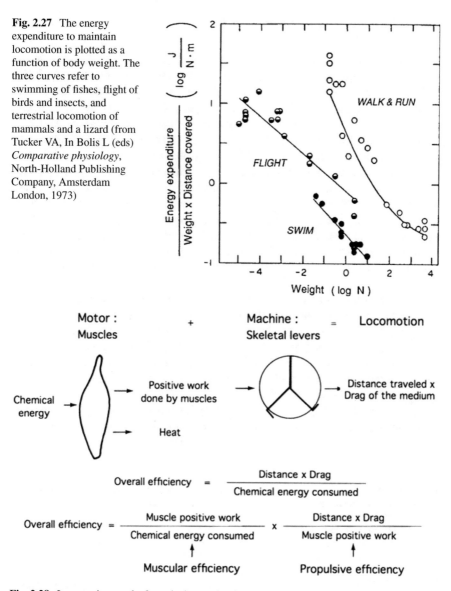

Fig. 2.28 Locomotion results from the interaction between a motor (the muscular system), which transform chemical energy into positive mechanical work (a very complex task, as we have seen, equal in frog and humans), and a machine, i.e. a lever system, which simply transforms force and displacement at its input into force and displacement of different size at its output (Fig. 2.16). As shown, the overall efficiency depends on the interaction between these two systems and is lower in legged terrestrial locomotion then in flying and swimming due to a different interaction of the machine with the surrounding (from Cavagna GA *Physiological aspects of legged terrestrial locomotion*, Springer International Publishing, 2017)

2.9.1 Internal and External Work

The problem of legged terrestrial locomotion is the *fixed point of contact of the foot on the ground*. As explained below this requires work to be done by muscles (Fig. 2.29).

During contact the forward velocity of the foot (or the feet) relative to the ground is nil. It is obvious however that, over a complete cycle of step, the average velocity of the foot relative to the ground must equal that of the center of mass of the whole body, otherwise a deformation of the body would take place. This means that the foot (and the whole lower limb) must be accelerated *relative* to the center of mass of the whole body in some phases of the step to maintain the same average forward velocity. These back-forward velocity changes of the limbs *relative* to the center of mass of the whole body imply *kinetic energy changes* within the whole body system. The work done to sustain these kinetic energy changes is called *internal work* $W_{int} = \Sigma 1/2\, m_i\, v_{ir}^2$, where v_{ir} is the velocity of the m_ith body mass segment *relative* to the center of gravity of the body. The internal work can be measured by a cinematographic procedure recording the displacements of the limbs relative to the center of mass as a function of time.

Contrary to the motion of a wheel where the link between its center of mass and point of contact on the ground is at each instant perpendicular to the velocity of the center of mass, during legged terrestrial locomotion, as in a 'square' wheel, the projection of the velocity vector on the link between center of mass and ground causes velocity changes and vertical displacements of the center of mass. The work done to sustain these kinetic and gravitational potential energy changes of the center of mass of the whole body is called *external work* (Fig. 2.30).

The external work W_{ext} can be measured from the force-time records obtained during locomotion of a subject on a *force-platform*. The general procedure is described in detail by Cavagna (*J. Appl. Physiol.* 39: 174, 1975).

Fig. 2.29 Left: back-forward movement of the limbs causes internal work. Right: interaction of the center of mass with the ground causes external work (from Cavagna GA *Physiological aspects of legged terrestrial locomotion*, Springer International Publishing, 2017)

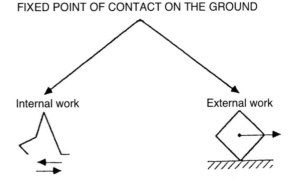

FIXED POINT OF CONTACT ON THE GROUND

Internal work   External work

Fig. 2.30 In a wheel, the velocity of the center of mass can be maintained constant because, at each instant, it is perpendicular to the link between center of mass and ground. On the contrary, in legged terrestrial locomotion, as in a 'square wheel', the component of the velocity on the constraint causes forward and vertical velocity changes (from Cavagna GA *Physiological aspects of legged terrestrial locomotion*, Springer International Publishing, 2017)

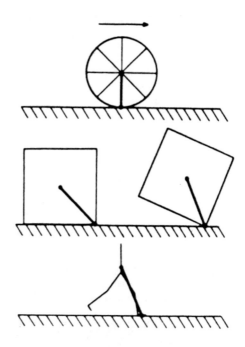

2.9.2 Walking

Nature faces the external work problem described above with two fundamental mechanisms of legged terrestrial locomotion: the *pendular mechanism of walking* at low and intermediate speeds and the *elastic mechanism of running* (hopping, trotting) at high speeds (Fig. 2.31). In the first case the link between center of gravity of the body and point of contact on the ground is maintained more or less rigid with the consequence that the center of mass pole vaults over the ground: the kinetic energy is transformed into gravitational potential energy until the highest point is reached and then back into kinetic energy of forward motion in the second phase of the step as in a pendulum (Fig. 2.32).

With increasing speed of locomotion the kinetic energy of forward motion lost during the first phase of the step is too large to be totally stored as gravitational potential energy because the vertical displacement at disposal is limited by the length of the lower limb (unless stilts are worn). In this case the link flexes absorbing mechanical energy during the first phase of the step and releasing mechanical energy during the second phase of the step. The mechanical energy (kinetic and gravitational) lost during the brake is transformed within the contracting muscles into heat and 'elastic' energy, which can be reutilized in the second phase of the step during the push. In spite of this partial storage-release of mechanical energy, running is more expensive than walking: 1 kcal/kg km in running and 0.5 kcal/kg km at the optimal speed of walking.

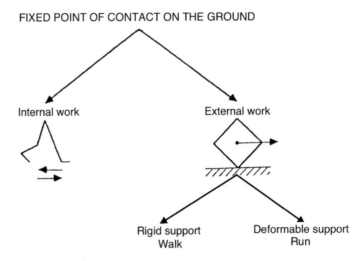

Fig. 2.31 The external work necessary to sustain the motion of the center of mass (right) is done according two fundamental mechanism: the pendular mechanism of walking (rigid support) and the bouncing mechanism of running (deformable support)

Fig. 2.32 Human and monkey (*Macaca speciosa*) walking over a force platform (0.5 × 4 m) used to determine the curves of kinetic energy, gravitational potential energy and total mechanical energy of the center of mass of the body during its forward and vertical motion (Fig. 2.33) (experiments made by Cavagna GA, Heglund NC and Taylor RC *Am. J. Physiol.*, 233: R243, 1977)

Figure 2.33 shows few steps of a walking human measured by integration of the force-platform signals. The motion of the center of gravity of the body during walking may be compared with that of a ball that falls and rises frictionless on a rigid support represented on top of the Figure. The kinetic energy E_{kf} and the potential energy E_p change in opposition of phase, i.e. a continuous exchange exists between kinetic and potential energy as indicated by the horizontal arrows in the schema at the bottom of the Figure. It follows that the positive external work actually done to sustain the total mechanical energy changes of the center of mass of the body, W_{ext}, is less than the sum of the positive work to sustain the velocity changes of the center of mass in forward direction, W_f, and of the positive work done against gravity, W_v. At the freely chosen speed, muscles perform external positive work in two phases of the step: in one phase (increment a, upward arrow at the bottom of the Figure) to increase the kinetic energy E_{kf} beyond the value attained thanks to the decrease of the potential energy E_p, and in another phase (increment b, upward arrow at the bottom of the Figure) to complete the lift of the center of mass beyond the value attained thanks to the decrease of the kinetic energy E_{kf}. The same quantity of mechanical energy ($a' + b'$) is absorbed by the muscles (downwards arrows) with the consequence that the total mechanical energy of the center of mass over the whole step cycle is unchanged. In other words, when the center of mass approaches the lowest point of its trajectory, the increment in E_{kf} is greater than the decrement of E_p, with the consequence that E_{cm} increases indicating that some energy is added to the system to accelerate forwards the center of mass (positive external work). The trend is reversed immediately after, during the first part of the lift of the center of mass: E_{kf} now decreases more than the simultaneous increase in E_p, and E_{cm} decreases indicating that some energy is subtracted from the oscillating system by braking the forward motion of the center of mass (negative external work): energy is added to be subsequently retrieved. When the center of mass approaches the highest point of its trajectory, the increment in E_p is greater than the decrement of E_{kf}, with the consequence that E_{cm} increases indicating that some energy is added to the system to complete the lift of the center of mass (positive external work). The trend is reversed immediately after, during the first part of the fall of the center of mass: E_p now decreases more than the simultaneous increase in E_{kf}, and E_{cm} decreases indicating that some energy is subtracted from the system by absorbing the excess of gravitational potential energy (negative external work). Again, energy is added and subsequently retrieved. The mismatch between the E_p and E_{kf} transduction therefore requires both positive and negative external work to be done by the muscles: for some unknown reasons energy is added and subsequently retrieved in two phases of the step. Evidently the shape and the phase shift of the E_p and E_{kf} curves and their relative amplitude, affects the changes in the total translational energy of the center of mass in the sagittal plane.

The pendular *recovery* of mechanical energy attained thanks to the transfer between E_p and E_{kf} can be defined as:

$$Recovery = \left(W_v + W_f - W_{ext}\right) / \left(W_v + W_f\right) = 1 - W_{ext} / \left(W_v + W_f\right)$$

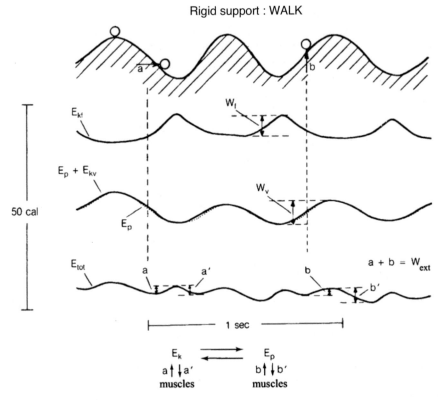

Fig. 2.33 Mechanical energy changes of the center of mass of the body during some steps of walking at 4.5 km/h of a 78 kg, 1.77 m and 23 years old man. The upper curve indicates the kinetic energy of forward motion, $E_{kf} = (M_b\, v_f^2)/2$ (where M_b is body mass and v_f the forward velocity of the center of mass). The middle continuous curve indicates the sum of the gravitational potential energy $E_p = M_b\, g\, s_v$ (g = acceleration of gravity, s_v = vertical displacement of the center of mass of the body) and of the kinetic energy of vertical motion, small during walking, $E_{kv} = (M_b\, v_v^2)/2$ (v_v = vertical velocity of the center of mass). The middle dotted curve, often not distinguishable from the continuous line, indicates the gravitational potential energy E_p only. The bottom curve indicates the total mechanical energy $E_{tot} = E_p + E_{kf} + E_{kv}$ (modified from Cavagna GA, Thys H, Zamboni A, *J. Physiol. London*, 262, 1976)

where W_v represents the positive work calculated from the sum, over one step, of the positive increments undergone by the gravitational potential energy $E_p = M_b$ $g\, s_v$ (where M_b is the mass of the body, g is the acceleration of gravity and s_v is the vertical displacement of the center of mass), W_f is the positive work calculated over one step, of the positive increments undergone by the kinetic energy of forward motion, $E_{kf} = (M_b v_f^2)/2$ (where v_f is the instantaneous forward velocity of the center of mass), and W_{ext} is the positive external work calculated from the sum, over one step, of the positive increments undergone by the total mechanical energy of the center of mass, $E_{tot} = E_p + E_{kf} + E_{kv}$. The kinetic energy of vertical motion

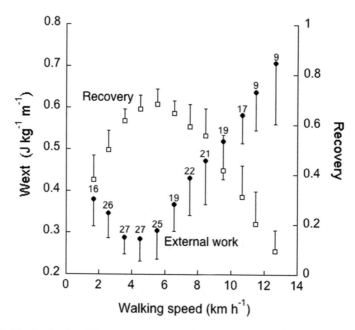

Fig. 2.34 The 'optimal' walking speed. Average and standard deviation (figures indicate the number of items in the mean) of the weight specific *external* mechanical work done per unit distance (filled circles), and of the *recovery* due to the transfer between gravitational potential energy and kinetic energy of forward motion of the center of mass (open squares). Other examples of this general mechanism are shown in Fig. 2.35

$E_{kv} = (M_b v_v^2)/2$ (where v_v is the instantaneous vertical velocity of the center of mass), has not been taken into account when obtaining the recovery from W_v and W_f. This is because E_{kv} has no effect on W_v since vertical velocity is zero at the top/bottom end points of the E_p curve.

In a frictionless pendulum the potential and kinetic energy curves would be the specular image of each other (i.e. $W_v = W_f$), their sum would be a horizontal line (i.e. $W_{ext} = 0$ and recovery $= 1$). Figure 2.33 shows that in walking, at intermediate speeds, a large exchange exists between kinetic and gravitational potential energy, but in spite of this, their sum is not an horizontal line, i.e. some external work has to be done to maintain the motion of the center of mass ($W_{ext} = a + b$).

Since a long time it is known that the metabolic energy expenditure to cover a given distance during walking attains a minimum at about 5 km/h. Figure 2.34 shows that the external work attains a minimum and the *recovery*, as defined above, attains a maximum of about 0.6 at about the same speed indicating that the energy expenditure is tightly bound to the pendular mechanism of walking.

As described above, with increasing speed the kinetic energy changes of forward motion become too large to be accommodated as gravitational potential energy and are absorbed by contracting muscles with the more expensive mechanism of running. Similarly on the Moon, where the gravitational potential energy is 1/6 that on Earth,

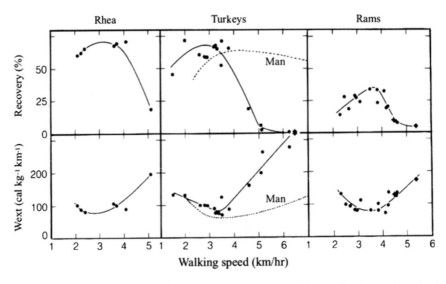

Fig. 2.35 As Fig. 2.34 to show that the pendular mechanism of walking applies also to other animal species (Cavagna GA, Heglund NC and Taylor RC *Am. J. Physiol.*, 233: R243, 1977)

Fig. 2.36 As in Fig. 2.34. Data on Mars obtained during ESA parabolic flight profiles (from Cavagna GA, Willems PA and Heglund NC *Nature*, 393: 636, 1998)

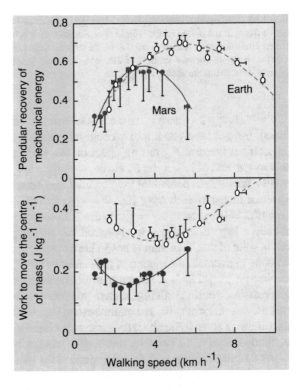

the exchange between gravitational potential energy and kinetic energy of forward motion is impaired (Margaria R and Cavagna GA, 1964) and, as shown by the pictures taken by the astronauts after landing, locomotion takes place through a succession of bounces with a mechanism similar to running. Also on Mars, where gravity is 0.4 g, the optimal walking speed and the velocity range attainable with the mechanism of walking are reduced (Fig. 2.36).

2.9.3   Running. Hopping and Trotting

The mechanism of running of humans and birds (Fig. 2.37, right) equals that of hopping (e.g. kangaroo, Fig. 2.37, left) and trotting of quadrupeds. In all these exercises the gravitational potential energy and the kinetic energy of forward motion of the center of mass change *in phase* during the step (Fig. 2.39), contrary to walking where, as we have seen, they change in opposition of phase (Fig. 2.33). The *pendular* mechanism of walking at low and intermediate speeds is substituted by the *bouncing* mechanism of running at high speed (Figs. 2.37 and 2.38).

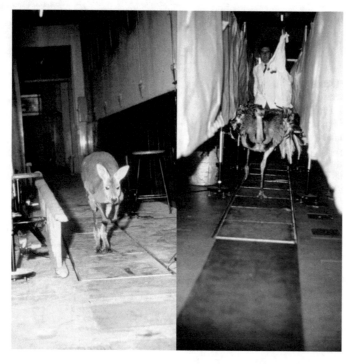

Fig. 2.37 Hopping kangaroo (*Megaleia rufa*, left) and running rhea (*Rhea americana*, encouraged by the author to run faster, right) on the force platform (experiments made by Cavagna GA, Heglund NC and Taylor RC *Am. J. Physiol.*, 233: R243, 1977)

Fig. 2.38 This cartoon shows how common sense may sometimes precede science. It was sent to us in the sixties by Professor Wallace Fenn whose fundamental work (Fenn WO *Am. J. Physiol.* 93, 1930b) has been our initial stimulus and guide. Here Fenn encouraged us to progress in the analysis of elastic storage and recovery in human running with its words "Keep the ball bouncing but beware of the boy"

Fig. 2.39 Mechanical energy changes of the center of mass of the body during a few steps of running at 12 km/h in man (59 kg, 1.72 m, 24 years) (modified from Cavagna GA, Thys H and Zamboni A, *J. Physiol. London*, 262: 639, 1976)

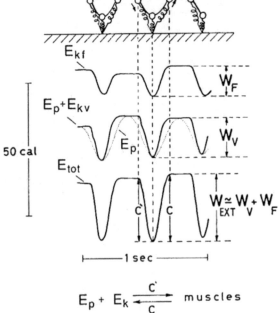

Figure 2.39 shows the curve of kinetic energy of forward motion $E_{kf} = (M_b v_f^2)/2$, whose amplitude is the work done W_f, the curve of gravitational potential energy $E_p = M_b g s_v$ (dotted line) plus the kinetic energy of vertical motion $E_{kv} = (M_b v_v^2)/2$, with amplitude W_v, and the curve of their sum $E_{tot} = E_p + E_{kf} + E_{kv}$, with amplitude W_{ext}. The horizontal tracts of the records indicate the aerial phase. Contrary to walking, the kinetic energy of vertical motion E_{kv} is appreciable in running: it attains a maximum before takeoff and it is transformed into gravitational potential energy during the aerial phase, when present (sometimes, as in trotting, the aerial phase may not take place). In all cases E_{kv} is nil at top and bottom of the oscillation and attains a maximum in between. The arrows at the bottom of Fig. 2.39 show that in running, contrary to walking (bottom schema in Fig. 2.33), both kinetic and gravitation potential energies are absorbed (c') and successively released (c) by contracting muscles (the stretch-shortening cycle). This is a very inefficient mechanism: contrary to flying and swimming where muscular work is done against frictions of air and water, in running, hopping and trotting, muscles work against themselves, the work done against air resistance is negligible and only a small fraction is recovered as 'elastic' energy; for this reason the weight specific energy expenditure per unit distance is greater than in walking at the optimal speed. The $Recovery = 1 - W_{ext}/(W_v + W_f)$ approaches zero in running because $(W_v + W_f) \approx W_{ext}$.

2.10 The Heart

As mentioned above, exchanges between external and internal environment are made possible by two mechanisms: diffusion for short distances (at cellular level) and circulation of blood for large distances (between different organs). Circulation is sustained by the rhythmic contraction of the heart: left ventricle for the systemic circulation and right ventricle for the pulmonary circulation. Muscular force developed by the heart performs solely *positive* work, which is finally degraded as heat by friction between blood layers (see Circulation).

2.10.1 Action Potential

Figure 2.40 shows a section of the heart. Note that the thickness of the walls of the right ventricle is smaller than that of the left ventricle because the pressure of blood in the pulmonary circulation is about 1/6 that in the systemic circulation. There are two kinds of cardiac cells: the *undifferentiated* muscular cells, which represent ~100% of the total and are responsible for the mechanical work done by the heart, and the *differentiated* or *pacemaker cells*; these last are responsible: (i) for the rhythmic spontaneous onset of the *action potential* at the level of the *sinus-atrial node* initiating each contraction of the heart; and (ii) for the *fast* invasion through the *atrial-ventricular* node and its branches of the undifferentiated cells of both ventricles resulting in their

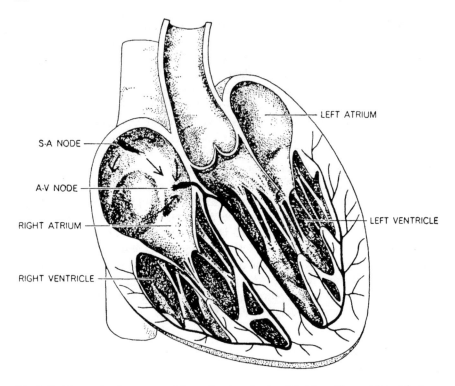

S-A NODE

A-V NODE

RIGHT ATRIUM

RIGHT VENTRICLE

LEFT ATRIUM

LEFT VENTRICLE

Fig. 2.40   The conduction system of the cardiac impulse (modified from Harary I *Scientific American* 206: 141, 1962)

simultaneous contraction. Both these characteristics are under control of the parasympathetic (inhibitory of heart frequency: *vagal tone*) and sympathetic nervous systems (excitatory). Whereas in skeletal striated muscles excitation is confined within each neuromotor unit, cardiac striated muscle behave as a *functional syncytium* resulting in a simultaneous contraction of all its parts that guarantees an effective expulsion of blood through the pulmonary artery and the aorta. Heart is surrounded by a *serous membrane called pericardium*, which has three functions: (i) allowing movement of the heart relative to the surrounding tissues, (ii) limit an excessive increase in heart volume, and (iii) being fixed to the diaphragm: this forces during ventricular contraction a lowering of the ventricular base, which facilitates influx of blood into the atria.

Three differences exist between skeletal and cardiac muscle: (i) the force exerted can be varied in skeletal muscle by recruiting a different number of neuromotor units, whereas in cardiac muscle the force is always the maximal exerted for a given level of parasympathetic and sympathetic tone; (ii) the *spontaneous onset* of the action potential in the heart (normally in the atrial-ventricular node, but in some conditions in other parts of the conduction system); (iii) the dependence of force from the frequency of stimulation in each neuromuscular unit of skeletal muscle

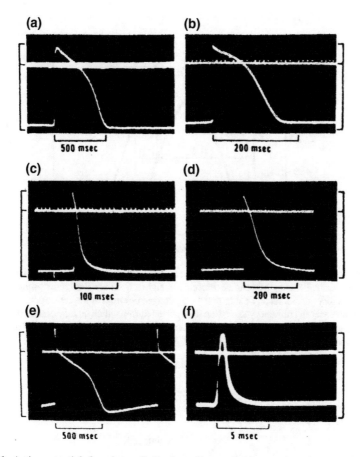

Fig. 2.41 Action potentials from intracellular electrode recorded from various tissues. **a** Frog heart. **b** Dog ventricle. **c** Rat ventricle. **d** Dog auricle. **e** Sheep Purkinje tissue, and, **f** Rat skeletal muscle for comparison. Ordinates voltage calibration (from above to below) 30 mV (positive inside the cell), 0 mV, 100 mV (inside negative). Note that the action potential is much faster in skeletal muscle than in cardiac muscle (from Weidmann S *Ann. New York Acad. Sc.* 65: 663, 1957)

(twitch, clonus and tetanus), whereas in cardiac muscle the force exerted is always that of a single twitch.

Several *action potential* of different cells are compared in Fig. 2.41. It can be seen that: (i) the electrical potential originally negative inside the cell becomes quickly positive during the action potential and reverts to its initial value much faster in skeletal muscle (*F*) than in the cardiac muscle where its duration approaches that of the force development during the mechanical contraction (Fig. 2.42); (ii) the negative intracellular potential, which is stable before the action potential in the *undifferentiated* muscular cells (cardiac and skeletal), drifts to the threshold starting the subsequent action potential in the *pacemaker cell* (*E*).

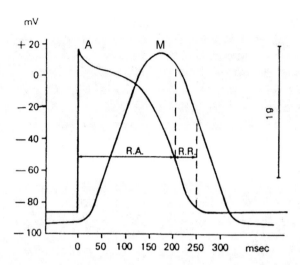

Fig. 2.42 Superposition of the action potential of an undifferentiated cardiac muscular cell (*A*, left) with the mechanical response of cardiac muscle (*M*, right). *R.A.*: absolute refractory period; *R.R.*: relative refractory period. It can be seen that the prolonged duration of the action potential (see also Fig. 2.41) increases the refractory excitation time period during which no other action potential and mechanical response can take place (modified from Keele CA and Neil E *Samson Wright's applied physiology*, Oxford University Press, London, 1971)

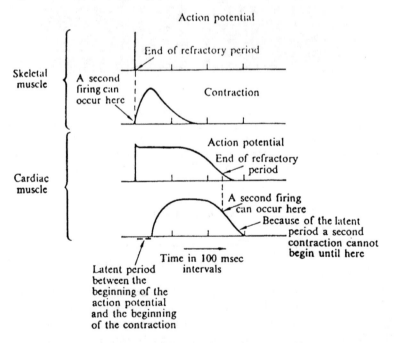

Fig. 2.43 A comparison of the action potentials and subsequent contractions in skeletal and cardiac muscle (from Horrobin DF *Essential physiology*, MTP Press Limited, Lancaster Boston The Hague Dordrecht, 1973)

Figure 2.43 shows an additional difference between skeletal and cardiac muscle: *whereas in skeletal muscle there is virtually no latent period between the action potential and the beginning of the contraction, in cardiac muscle the latent period is relatively long.* If a firing occurs immediately after the end of refractory period an *extrasystole* occurs followed by a longer silent period till the next rhythmic firing takes place (compensatory pause).

2.10.2 Pressure-Volume Changes

Figure 2.44 shows the pressure and volume changes in the cardiac cavities, in the aorta and in the pulmonary artery as a function of time during a heart cycle. Note that the trend of the pressure changes are similar in the left and right ventricles, but that the arterial pressure in the right ventricle is about 1/6 that in the left ventricle; the volume changes are the same in left and right side of the heart. In what follows we will describe the left side of the heart only.

The pressure base line is set equal to zero for simplicity (actually since the heart is in the mediastinum between the two lungs, the pressure during the time before contraction depends on air pressure within the lungs and the tendency of lungs to collapse as we will see later on). In the first 0.1 s on the abscissa of Fig. 2.44, corresponding to the end of the diastole, the ventricular volume attains 100% thanks to the atrial systole. When the ventricle starts contracting, its inside pressure increases, but its volume remains constant for about 50 ms until the minimal pressure in the aorta is reached (~80 mm Hg); in this *isometric contraction period* both the *mitral* and *aortic* valves are closed. After the isometric period, the pressure in the aorta and in the ventricle increase together for about 150 ms until the maximal pressure of 120 mmHg is attained invading a tract of the arterial tree of about 2 m in the adult at rest (see the end of Sect. 1.5.2). The increase in pressure takes place during the period of maximal efflux as shown by the ventricular volume record in Fig. 2.44 (when the input of blood from ventricle into the arterial tree is greater than the output towards the veins through the PRU). The volume of the ventricle varies from a maximum of ~160 ml to a minimum of ~80 ml. As shown in Fig. 2.44 about 80% of blood volume leaves the heart during the period of maximal efflux during the rise of pressure in the ventricle (and in the aorta). After the periods of maximal and reduced efflux, the ventricle relaxes and its inside pressure falls. Blood flows from the atrium into the ventricle after the *isometric relaxation period* (50 ms in Fig. 2.44), during which the volume of the ventricle remains constant. During this period both the *aortic valve* (between ventricle and aorta) and the *mitral valve* (between atrium and ventricle) are closed. After this period the mitral valve opens and blood flows from the atrium into ventricle with the consequence that the volume of ventricle increases first more rapidly and then more slowly platooning at ~70% of the stroke volume (Fig. 2.44). At this point the atria contract completing the ventricular filling causing a small increase in pressure in the atrium-ventricular cavity (atrial systole) and a new cardiac cycle begins (from 0 to 0.1 s in Fig. 2.44). Contraction of the atria is not essential for life:

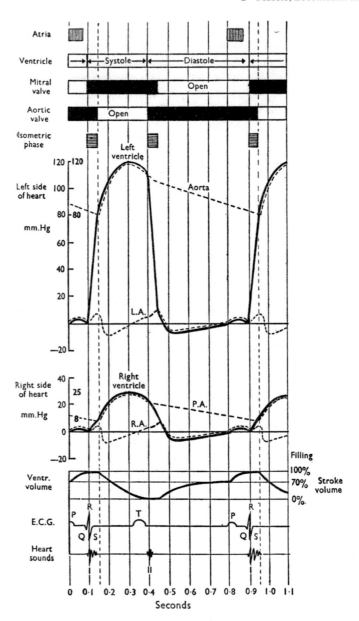

Fig. 2.44 The sequence of events in the cardiac cycle starting with the onset of the atrial systole. Part of the subsequent cycle is shown. P.A. = Blood pressure in the pulmonary artery; R.A. = right atrial pressure; L.A. = left atrial pressure. The electrocardiogram waves (corresponding to the depolarization of the atria, P, the depolarization of the ventricles, QRS, and the repolarization of the ventricles, T) are also shown together with the heart sounds (I: closing of the atria-ventricular valves, II: closing of the semilunar valves) (from Green JH An introduction to human physiology, Oxford University Press, London, 1968)

in atrial fibrillation (when the walls of the atrium oscillate as a bag of worms due to a sparse excitation of their muscular cells) the activation of the ventricles is arrhythmic due to the irregular input of action potentials to the atrium-ventricular node. Whereas during the ventricular *systole* the pressures in the aorta and in the left ventricle are equal (interrupted and continuous lines in Fig. 2.44), at the end of systole the two curves diverge thanks to the closure of the *aortic valve*: the pressure in the ventricle falls quickly because ventricle relaxes and the pressure in the aorta falls slowly as blood flows into the arterial tree.

Let's now consider the pressure in the atrium. The record in Fig. 2.44 initiates with a positive wave due to atrial systole; this is followed by another positive wave due to a partial upward extroflexion of the closed *mitral* valve taking place during the *isometric* contraction of the ventricle. When the aortic valve opens and blood exits from the ventricle, the basis of the heart lowers (since the apex is fixed to the diaphragm) and this causes a fall of pressure in the atrium due to its increase in volume. *It is important to realize that the flow of venous blood into the heart is continuous.* During the systole, when the mitral valve between atrium and ventricle is closed, this continuous flow causes a rapid increase of the pressure in the atrium cavity only. After the systole, during the diastole, when the mitral valve is open, the blood accumulated in the atrium discharges suddenly in the ventricle, the pressure in the atrium falls (from 0.45 to 0.5 s in Fig. 2.44), but after this, during the diastole it keeps increasing even if less steeply than before due to the continuous flow of venous blood into the larger atrium-ventricular cavity. At the end of the diastole, atria contracts and the cycle repeats itself.

Figure 2.45 shows, together with the heart sounds due to closing of the atrial ventricular valves (I) and the semilunar valves (II), which are easily audible with a phonendoscope, the venous jugular pulse which is not audible due to the small values of the venous blood pressure. The trend of the jugular pulse reflects that of the right atrial pressure changes illustrated in Fig. 2.44. The *a* wave corresponds to the right atrial contraction. The *c* wave corresponds to right ventricular contraction causing the tricuspid valve to bulge towards the right atrium. The descent of pressure to *x* occurs as a result of the right ventricle pulling the tricuspid valve down during right ventricular systole. The *d-v* wave corresponds to venous filling when the tricuspid wave is closed. The *y* descent corresponds to the rapid emptying of the atrium into the ventricle following the opening of the tricuspid valve. The following increase to *a* corresponds to the filling of the atrium-ventricular cavity by the input of the continuous venous flow.

The cardiac tones are due to the *closure* (as when you clap your hands) of the atrium-ventricular valves and the semilunar valves. The *first tone is due to the closure of the atrium-ventricular valves*; the *second tone is due to the closure of the semilunar valves*. The first tone is more prolonged and has a lower frequency than the second tone. The first tone lasts ~0.15 s and has a frequency of ~25 Hz. The second tone lasts ~0.1 s and has a frequency of ~50 Hz. Figure 2.46 shows the auscultatory areas on the chest of the tones due to closure of: (i) semilunar valves: *aortic* (*A*) and *pulmonary* (*P*); (ii) atrium-ventricular valves: *tricuspid* (*T*), between *right* atrium and *right* ventricle, and *mitral* (or *bicuspid*) (*M*) between *left* atrium and *left* ventricle. Two

Fig. 2.45 Above: heart sounds. Below: venous pulsations recorded from the right external jugular vein (from Margaria R and De Caro L *Principi di Fisiologia Umana*, Vallardi, Milano, Italy, 1977)

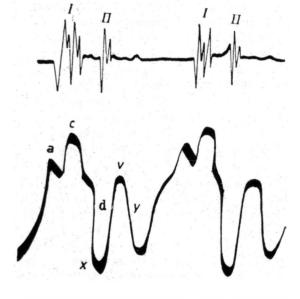

Fig. 2.46 Auscultatory areas on the chest of the cardiac tones. Semilunar valves: *aortic* (*A*) and *pulmonary* (*P*). Atrium-ventricular valves: *tricuspid* (*T*) and *mitral* (*M*) (from Burton AC *Physiology and biophysics of the Circulation*, Year Book Medical Publisher Inc., 1968)

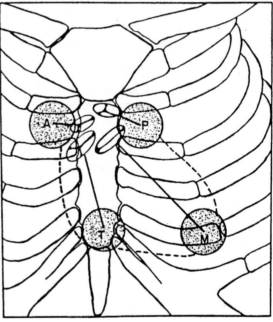

other tones are barely audible in some subjects: a *third* tone about 0.2 s after the second tone apparently due to the passage of blood from the atria to ventricles and a *fourth* tone, just before the first tone, due to atrial systole.

In pathological conditions a prolonged duration of the first tone may be due to turbulence of blood flow due to an *insufficiency* of the atrium-ventricular valves (some blood flows back into the atrium during the contraction of ventricle) or to a *stenosis* at the level of the semilunar valves. On the other hand a prolonged duration of the second tone may be due to and *insufficiency* of the semilunar valves (some blood flows back into the ventricle) or to a stenosis of the atrium-ventricular valves. In addition to differences in sound production, insufficiency and stenosis of the atrium-ventricular and semilunar valves cause a deformation of the sphygmic wave (Fig. 1.22) and of the volume of ventricles and atria.

2.10.3   The Starling Law

As described in Sect. 1.5.2 and Fig. 1.26, the blood output of the left ventricle *must equal* the blood output of the right ventricle. Since it is practically impossible to get two pumps with the *exactly* the same output, *Nature admits the error, but provides a mechanism capable to correct it: the Starling law*, which consists in a *negative feedback*. When the volume of a ventricle increases because it receives more blood than it has emitted, its contraction and blood output become more effective thus compensating the initial excess in volume. This mechanism can be described by a graph having on the ordinate the work done by ventricular contraction (Work = average pressure × volume emitted = $\bar{P} \times \Delta V$) and on the abscissa the volume of the heart cavities (atrium plus ventricle) before the systole (telediastolic). Since this volume is difficult to measure and since the greater the volume the greater is the elastic distension of the 'parallel' elastic elements in the heart walls, it is convenient to set on the abscissa the telediastolic *pressure* as an index of the ventricular volume before the systole (Fig. 2.47). According to the law of Starling, the greater the volume, the greater should be the work done during the systole to compensate the excess distension of the heart cavities with a greater blood output. The Starling law in vivo is affected by: (i) the law of Laplace, (ii) the cardiac *troponin C* and (iii) the sympathetic and parasympathetic neurovegetative tone.

As described in Chap. 1, paragraph 1.6.1 The Laplace law applied to a sphere (assuming cardiac cavities a spherical cavity) is

$$P = (1/R + 1/R)T = 2T/R$$

where R are two arcs perpendicular to each other in each point of the surface. It follows that the *active* tension that the ventricle must develop is $T_a = PR/2$. The active tension, in turn, is a force F acting perpendicularly to the circumference, i.e. $T_a = F/(2\pi R)$. It follows that $F/(2\pi R) = PR/2$, i.e. $F = \pi PR^2$ *showing that, for a given pressure P, the force F that the cardiac muscle cells must develop increases with the square of the radius R of the heart!* It follows that, for a given physiological condition, a small heart is favored relative to a large heart. We should realize that an enlargement of the heart in a patient requires more force to be developed by his

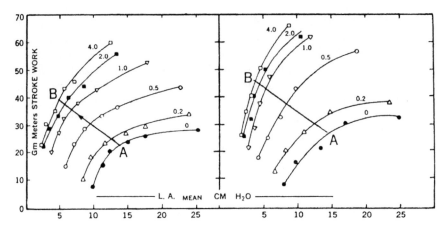

Fig. 2.47 Abscissa: average telediastolic ventricular pressure as an indication of the distension of the heart. Ordinate: work done during the systole by stimulation of the cardiac *sympathetic* nerves at the indicated frequencies (0–4 cycles per second). In the intact animal the operating point may move from *A* at rest to *B* in heavy exercise and so provide an apparent contradiction to Starling's law (modified from Sarnoff SJ and Mitchell JH *The control of the function of the heart* in Hamilton WF and Dow P (eds) *Handbook of Physiology* Sect. 2, Vol. 1 Circulation, American Physiological Society, Washington DC, 1962)

cardiac cells, i.e. more maintenance heat and greater energy expenditure. The Laplace law therefore has a negative effect on the Starling law, i.e. to compensate with a greater work an occasionally greater blood input into the ventricle. This compensation therefore must rely on the isometric force-length diagram of the individual muscular cells, i.e. that cross bridge action is more effective *up to a limit* with the increase of cell volume. As shown in Fig. 2.15 the force a sarcomere can develop increases with length to a maximum because, with increasing length, decreases the interference due to the overlap between actin filaments within the sarcomere. This is not the case for the increase in force with sarcomere length in the heart cells. The increase in force with length, compensating an excess in the heart volume, is due to a particular molecule the *cardiac troponin C*, which releases more calcium ions when lengthened, resulting in a greater displacement of the inhibitory troponin I and in a greater number of attached cross-bridges. Above a given limit however both the Laplace law and the force-length diagram will lead to decompensation, i.e. to heart failure.

When measured on the living animals, instead of on the heart-lung preparation as Starling did, scattered results were obtained, some apparently in contrast with the Starling law. This was due to the fact that in the integer animal the heart is subjected to the influence of the *positive inotropic effect of the ortosympathetic autonomic nervous system* increasing the force and the mechanical power of heart contraction (let's remember that the parasympathetic autonomic nervous system instead has mainly the effect to reduce heart frequency). Figure 2.47 shows *isofrequency ortosympathetic stimulation work-volume* curves (volume is taken on the abscissa as an increasing function of the telediastolic relaxed muscle left atrium pressure). It can be seen that at a given frequency of ortosympathetic stimulation the heart compensate with a greater positive work done during the systole an increase of its initial volume in agreement with the Starling law. The slope of the curves decreases with volume indicating that compensation attains a limit after which heart failure may occur. The A–B lines show that work may increase when the volume is decreased, in contrast with the law of Starling, when ortosympathetic stimulation is not taken into account.

2.10.4 Work

The positive work done by the heart is dissipated into heat by friction between blood layers particularly near the wall of the vessels, not against the wall of the vessels (Fig. 1.11). Mathematically the work done is $W = \bar{P}_{ventricle} \times \Delta V$ where $\bar{P}_{ventricle}$ is the *average* pressure in the ventricle and ΔV is the volume of blood emitted (80 ml in Fig. 2.48). In the aorta some of the work done is found as kinetic energy, i.e.: $W = \bar{P}_{aorta} \times \Delta V + 1/2\,\delta v^2 \Delta V = \Delta V \times (\bar{P}_{aorta} + 1/2\,\delta v^2)$ where v is the average velocity of the blood in the aorta, δ is blood density and $\delta\,\Delta V$ is the mass of the volume emitted.

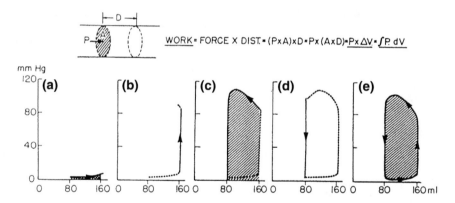

Fig. 2.48 Pressure (in mmHg) is plotted as a function of volume (in milliliters) of the human left ventricle during a complete diastolic-systolic cycle. **a** Filling of the relaxed ventricle during the diastole; the shadowed area below the curve represents the work done by stretching the parallel elastic elements during ventricle filling. **b** Pressure increase during ventricle isometric contraction until the minimal pressure in the aorta is attained (Fig. 2.44). **c** The volume of the left ventricle decreases during the systole; the pressure increases during the period of maximum efflux and subsequently decreases during the period of reduced efflux (Fig. 2.44); the dashed area below the curve represents the positive work done during the systole. **d** Pressure decrease during ventricle isometric relaxation after closure of the aortic semilunar valve. **e** The shaded area represents the net positive work done by the *active* muscular force during the systole assuming no *stress-relaxation*, i.e. no hysteresis in the recoil of the parallel elastic elements (From Burton AC *Physiology and biophysics of the circulation*, Year Book Medical Publisher Incorporated, Chicago, 1965)

2.10.5 Cardiac Output

The *cardiac output* is a *flow* ($l^3\,t^{-1}$) = *stroke volume* (volume emitted at each cardiac pulse, l^3) × heart frequency (t^{-1}); for example the cardiac output *at rest* could be 4800 ml (milliliters)/min = 80 ml (cc, cubic centimeters) × 60 beats/min. The cardiac output changes in several physiological conditions (sleep, strong emotions, digestion etc.), but it attains maximum during a maximal physical exercise to allow oxygen input and carbon dioxide output of the contracting muscles. *On average the increase in cardiac output takes place thanks to an increase of the heart frequency with a constant stroke volume* (Fig. 2.49).

The resting oxygen consumption is ~250 ml/min (the brain oxygen consumption is always the same independent of cerebral activity). The constancy of the stroke volume with increasing intensity of the muscular exercise (Fig. 2.49) is an *active* natural phenomenon; if heart frequency is artificially increased in a resting patient with a pacemaker, the stroke volume decreases because the time at disposal to fill the heart is reduced. During muscular exercise the stroke volume is maintained constant by two mechanisms: (i) the *muscular pump*: the contracting muscles squeeze the venous vessels, which, being provided by valves, increase the flow of venous blood towards the heart increasing its filling in spite of the less time at disposal; (ii) the stimulation of the cardiac sympathetic nerves, which increase the force of contraction of the ventricles during the systole: the *positive inotropic effect* on the *power* developed during ventricular contraction (Fig. 2.47). The limiting factor to the maximal oxygen

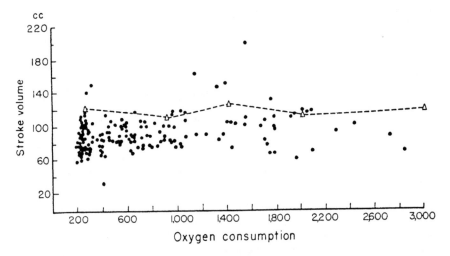

Fig. 2.49 The stroke volume (in milliliter = cc) is plotted as a function of the oxygen consumption from rest (~250 ml/min) to maximal physical exercise (~3000 ml/min, i.e. 3 L/min). The interrupted line refers to an athlete. Note that the number of data decreases with the oxygen consumption, i.e. with the capability to attain high intensity of the exercise in aerobic conditions (adapted from Rushmer RF and Smith OA Jr. *Cardiac Control Physiol. Rev.* 39: 41, 1959)

consumption, and therefore to the maximal *aerobic* physical exercise that can be maintained at will, is the flow of blood due the limited heart power output, not the flow of air to ventilate the lungs, which is always in excess. The maximal heart frequency attained during maximal exercise is the same in a normal subject and in a athlete: ~180 beats/min; the difference between the two is the cardiac frequency at rest, which in a athlete can be as low as 40 beats/min requiring a stroke volume of 120 ml to get the ~5 l/min of the cardiac output at rest (triangles in Fig. 2.49), as compared with the 60 beats/min with a stroke volume of 80 ml in the normal subject (filled points in Fig. 2.49). It follows that the maximal cardiac output in an athlete will be 120 ml × 180 beats/min = 21.6 l/min, whereas in a normal subject will be 80 ml × 180 beats/min = 14.4 l/min.

A lower frequency has the beneficial effect to increase the resting period of the heart relative to the work duration. At the heart frequency of 40 beats/min the period of ventricular activity is 30% of the total cycle and the period of rest during the ventricular diastole is 70% of the total, whereas the contrary is true at the heart frequency of 180 beats/min. This is particularly important because during the diastole the coronary blood flow is at a maximum.

As described previously (Sect. 2.8.2), the *efficiency of muscle* $= Work/M + ax + \bar{F}x + recovery\ heat = \bar{F}x/2(M + ax + \bar{F}x) = 0.20-0.25$. It was stressed that the maintenance heat M is much greater than the sum of shortening heat ax and the work done $\bar{F}x$. Correspondingly:

$$efficiency\ of\ heart = \bar{P}\Delta V/2(A + a'\Delta V + \bar{P}\Delta V)$$
$$= \bar{P}\Delta V/2[A + \Delta V(a' + \bar{P})]$$

where $\bar{P}$ is the average arterial pressure, ΔV is the stroke volume and A is the activation heat (since in the heart we do not have a tetanus, i.e. a maintenance heat $M \sim \Sigma A$). Of the two terms defining the *Work*, the average pressure $\bar{P}$ requires an energy expenditure much greater than the stroke volume ΔV. This means that in a cardiac patient it is much safer to increase the work of the heart by increasing the term ΔV (e.g. by decreasing the PRU as in a slow walk) than by increasing the term $\bar{P}$ (e.g. in a quarrel, anxiety or a strong emotion, when the PRU increase). Increasing the heart work thanks to an increase in ΔV, which has a little effect on the denominator (where $\bar{P}$ and A are the largest terms), results in an increased *efficiency* because the numerator increases relatively much more than the denominator (e.g. 1/100 vs. 2/102: the efficiency ~doubles).

An increase in frequency is also dangerous, because the A term is large and, as described above, the ratio between the period of ventricular activity A and the period of resting ventricular diastole increases with heart frequency.

In addition we must remember that the hydrostatic indifferent point in the erect position is set just below the heart, that the volume of the heart increases when the subject is laying down and that the force that the cardiac muscle must develop increases with the square of the radius according to the Law of Laplace. It follows that it is better to keep a cardio patient sitting instead of lying horizontally on the bed.

2.10.6 Measurement of Cardiac Output

There are two methods for measuring the cardiac output: the *Fick Principle* and the *Dyes Method*.

The *Fick Principle* states that the oxygen consumption in unit time equals the volume of oxygen concentration in its arterial blood minus the volume of oxygen in the same volume of its venous blood times the volume of blood flowing from aorta to the right atrium in the same time: $\dot{V}O_2 = {}_A\Delta_V \times \dot{V}_{blood}$, where $\dot{V}O_2$ is the oxygen consumption in unit time, ${}_A\Delta_V$ is the difference in oxygen concentration between aorta arterial blood and mixed venous blood and $\dot{V}_{blood}$ is the cardiac output. It follows that the cardiac output at rest is $\dot{V}_{blood} = \dot{V}O_2/{}_A\Delta_V$, i.e. 250 ml O_2 of oxygen consumption/min divided by 50 ml O_2 released by one liter of blood (Fig. 3.51) equals a cardiac output $\dot{V}_{blood} = 5$ l/min. Whereas the oxygen consumption can easily be measured by means of a *spirometer,* and the oxygen content in the arterial blood can be measured by an arterial blood sample (with a *Van Slike* apparatus), the oxygen concentration of the *mixed venous blood* in the right atrium or in the pulmonary artery is obviously more difficult to measure.

The Stewart-Hamilton *Dye Method* allows to overcome this difficulty. This method consists in injecting a given volume of dye in any vein and record the subsequent dye wave passing through the earlobe by means of a photocell. In order to understand this method, let's consider the following drawing.

The quantity of dye injected is $Q = \bar{C} \times V_{blood}$ where $\bar{C}$ is the *average* concentration of dye in blood after mixing with the *total* blood volume, V_{blood}, at the level of the heart (Fig. 2.50). As described before (Sect. 1.5 and Fig. 1.21, defining why the average pressure can be expressed as $\bar{P} = \int p dt/\tau$), we understand that $\bar{C} = \int c dt/\Delta t$ where $\int c dt$ is the area below the *primary curve* in Fig. 2.51, i.e. the dye wave lasting the time interval Δt (about 20 s in Fig. 2.51). Since $V_{blood} = \dot{V}_{blood} \times \Delta t$, the equation $Q = \bar{C} \times V_{blood}$ can be written $Q = (\int c dt/\Delta t) \times (\dot{V}_{blood} \times \Delta t)$, i.e. the cardiac output $\dot{V}_{blood}$ can be measured as $\dot{V}_{blood} = Q/\int c dt$. Note that, as shown in Fig. 2.51, the primary curve precedes the subsequent recirculation of polluted blood deriving from shorter circulation paths upstream the point of observation (e.g. the coronary circulation).

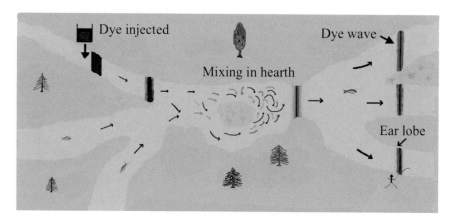

Fig. 2.50 A known quantity of dye is poured in a branch of a river (one vein of our body). It travels towards an island (the heart) where all the branches mix resulting in a single dye wave equal in all subsequent branches of the river. A fisherman will observe an increase, a maximum and a decrease of dye in front of him similarly to the record in Fig. 2.51 showing dye wave passing through the ear lobe

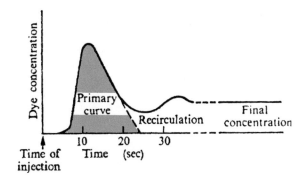

Fig. 2.51 Dye wave visible in any capillary district of the body (e.g. ear lobe) after injection of a dye into a vein. The gray area represents the true dye wave to be considered, preceding the remaining concentration of the injected dye into the blood. Since the left ventricle ejects at each systole half of the dye it contains, the fall in dye concentration follows an exponential decay resulting in a linear fall in concentration easily extrapolated when plotted in semi-log paper (redrawn from Moran-Campbell EJ, Dickinson CJ and Slater JDH (eds) *Clinical Physiology*, Blackwell, Oxford, 1963)

Chapter 3
Respiration

Abstract Consequences of the bidirectional flow of air within lung structures: dead space definition, composition of ambient air, exhaled air and alveolar air. The Law of Dalton: partial pressure of each component in a gas mixture. Dynamics of respiration: laminar and turbulent flow. The respiratory muscles. Effect of air motion and lung volume on the airway size. Measurement of lung volumes: spirometers of Benedict, Tissot and body plethysmograph. Flowmeter. Divisions of total lung capacity. Metabolism: basal and energetic. Methods of measurement of metabolism. The oxygen debt. Statics of the respiratory system. Pressure-volume diagram of living lung: area-surface tension diagram. Pulmonary hysteresis. Intrathoracic and intrapleuric pressure. Pressure-volume diagram of the combined system lungs-chest wall: the relaxation curve. Maximal expiratory and inspiratory pressures. Graphic and arithmetic representation of the respiratory work: optimal breathing frequency. Diffusion of gases: derivation of the first Fick's law. Resistances to diffusion: gas-gas, through tissues from air to hemoglobin, limited velocity to hemoglobin oxygenation and backpressure. Breathing at high altitude: maximal altitude attainable by humans. Equation of alveolar air. Diving in apnea and with a respirator. Pneumothorax. Ventilation-perfusion ratio. Gas transport in blood: oxygen and carbon dioxide. Acid-base equilibrium.

3.1 Introduction

Respiration is one of the ways where exchanges occur between the outside surrounding and the internal milieu of our body, in particular gas exchanges represented by oxygen O_2 input and carbon dioxide CO_2 output. In circulation the exchanges are provided by ventricular contractions for large distances, causing blood flow within the blood vessels, and by diffusion for the small distances at the capillary-cellular level. In respiration they are sustained by the action of the respiratory muscles for large distances, causing airflow within the airways, and by diffusion for small distances, at the capillary-alveoli level. In addition whereas in circulation the motion of blood is one way only thanks to valves, in respiration the motion of the air is two ways.

© Springer Nature Switzerland AG 2019

G. Cavagna, *Fundamentals of Human Physiology*,

https://doi.org/10.1007/978-3-030-19404-8_3

Figure 3.1 shows that $\Delta V_{lungs} = \Delta V_{chest\ wall}$, because the intrapleural fluid is incompressible; when $\Delta V_{chest\ wall}$ increases during inspiration (the piston in Fig. 3.1 moves to the left) the pressure within the lungs becomes lower than atmospheric pressure and air flows within the lungs, vice versa during expiration. The exchanges between air and blood take place at the level of the alveoli; the total surface of all the alveoli is about seventy square meters and the volume of blood spread over this large surface is only about hundred cubic centimeters! This allows rapid gas exchanges through a small distance by the mechanism of diffusion, but requires thicker airways to protect the thin alveolar walls from mechanical injuries. The volume of air contained in the thicker airways where the gas exchanges cannot occur is called *anatomical dead space and in the adult equals about* 150 ml.

These two characteristics of lung mechanics, *bi-directionality* and *dead space*, imply the formation of three gaseous environments with different composition in their gas mixture: *external or ambient air, exhaled air* and *alveolar air*. The composition can be expressed as *percentage (%)* or as a *fraction*.

The composition of the *ambient* air is given in Table 3.1.

Note that the composition of the ambient, inspired air is always the same *independent of the quote*, at sea level as on the top of the Everest due to its continuous atmospheric mixing (the remaining are rare gases: neon, krypton and xenon).

When the piston in Fig. 3.1 moves to the left, a volume of air, called *tidal volume* enters the lungs, V_{ti}. The tidal volume exhaled is V_{te}. Both these volumes at rest in an adult are $V_t \sim 500$ ml. When 500 ml of V_{ti}, are inspired 350 ml enter the alveoli and are subjected to CO_2 and O_2 exchanges whereas 150 ml *stay in the dead space and maintain the composition of the ambient air saturated with water vapor*. During

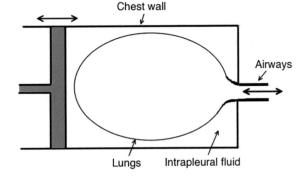

Fig. 3.1 The movement of the chest wall (ribs, diaphragm etc.) are simulated here as the movement of a piston to show that the air flow is bi-directional and that the volume changes of the lungs equal those imposed by the chest wall since the intrapleural fluid is incompressible

Table 3.1 Gas content in the ambient air

Gas	Percentage (%)	Fraction
Oxygen (O_2)	20.93	0.2093
Carbon dioxide (CO_2)	0.03	0.0003
Nitrogen (N_2)	79.03	0.7903
Total	99.99	0.9999

Table 3.2 Gas content in the exhaled air

Gas	Percentage (%)	Fraction
Oxygen (O_2)	16	0.16
Carbon dioxide (CO_2)	4	0.04
Nitrogen (N_2)	80	0.80
Total	100	1

Table 3.3 Gas content in the alveolar air

Gas	Percentage (%)	Fraction
Oxygen (O_2)	14	0.14
Carbon dioxide (CO_2)	6	0.06
Nitrogen (N_2)	80	0.80
Total	100	1

the following expiration a mixing of alveolar and dead space air takes place in the exhaled air; at the end of the expiration the composition of the exhaled air equals that of the alveolar air (see Fig. 3.15).

The *composition of the exhaled air*, resulting from the *mixing of the alveolar and dead space* air, is given in Table 3.2 (the following are reasonable easily memorized values which can vary in different respiratory conditions). The composition of the *alveolar air* is given in Table 3.3.

The question arises: why the % of nitrogen changes from 79%, in the ambient inspired air, to 80% in exhaled and alveolar air even if N_2 is not absorbed as O_2 nor produced as CO_2? The N_2 is an inert gas, which does not participate to the respiratory exchanges, i.e. its volume does not change. It follows that its percentage will increase if the total volume of the respiratory gases O_2 plus CO_2 decreases, vice versa if the total volume increases. If the volume of CO_2 added equals the volume of O_2 absorbed, the total volume of the mixture and the % of N_2 will not change. The ratio between volume of CO_2 added and volume of O_2 absorbed is called *respiratory quotient*: $QR = CO_2/O_2$. If $QR = 1$ the percentage of N_2 will equal that in the inspired air (79%). If $QR > 1$ the percentage of N_2 will decrease below that in the inspired air. If $QR < 1$ the percentage of N_2 will increase above that in the inspired air.

3.2 The Law of Dalton

The law of Dalton states that in a mixture of gases the *partial pressure exerted by each gas equals the pressure that that gas would exert if alone would occupy the whole volume of the mixture*. Suppose to divide the classroom into two equal halves, both of them in equilibrium with the external atmospheric pressure P_b; let's assume that one half of the classroom contains pure oxygen and the other half pure nitrogen. Since both halves are in equilibrium with the outside pressure and between them, each compartment will have the same pressure, i.e. $P_{O_2} = P_{N_2} = P_b$. Let's now mix oxygen and nitrogen within the classroom, so that each gas occupies the whole volume of the classroom. It is clear that in this case the distance between O_2 and CO_2 molecules will double, the fraction of each gas in the whole mixture will half and with it its partial pressure, i.e. $P_{O_2} = 1/2\,P_b = P_b \times F_{O_2}$ and $P_{N_2} = 1/2P_b = P_b \times F_{N_2}$ where F_{O_2} and F_{N_2} *are the fractions of oxygen and nitrogen content in the mixture* and P_{O_2} and P_{N_2} are the *partial pressures* of oxygen and nitrogen. Water molecules exist in addition to nitrogen, oxygen and carbon dioxide in the ambient, alveolar and expired air. The partial pressure of water vapor, P_{H_2O}, depends on its *saturation* and *temperature*. Saturation is 100% when water vapor is in equilibrium with its liquid phase as in the alveoli; its partial pressure increases with temperature as shown in Fig. 3.2. It can be seen that the partial pressure of saturated vapor of water at body temperature (~37 °C) is 47 mm Hg.

It follows that the partial pressure of oxygen in the alveoli will be: $P_{AO_2} = (P_b - 47) \times F_{AO_2} = (760 - 47) \times 0.14 = 100$ mmHg. The partial pressure of carbon dioxide in the alveoli and in the dead space at the end of expiration will be $P_{ACO_2} = (P_b - 47) \times F_{ACO_2} = (760 - 47) \times 0.06 = 43$ mmHg. I leave as an exercise the calculation of P_{ACO_2} and P_{AO_2} in the inspired dry air and in the dead space at the end of inspiration assuming $P_b = 750$ mmHg.

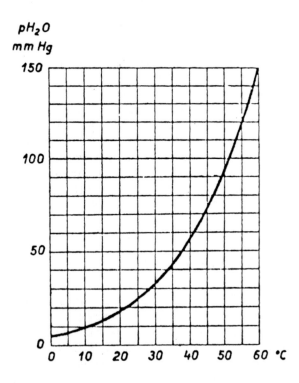

Fig. 3.2 Pressure of saturated water vapor in millimeters of mercury as a function of its temperature in degrees centigrade (from Margaria R and De Caro L *Principi di Fisiologia Umana*, Francesco Vallardi, Milano, 1977)

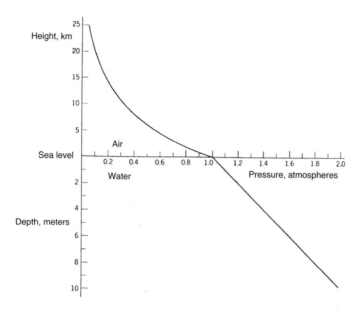

Fig. 3.3 Change of air pressure with altitude above sea level and water pressure with depth below sea level. *Note* (i) the different scale difference on the ordinates for air and water, and (ii) the pressure changes exponentially with height in the compressible air and linearly in the incompressible water

3.3 Effect of the Altitude

Since air (contrary to water) is compressible, at low altitudes it is denser than at high altitudes due to the weight of the column of air above it. The atmospheric pressure decays with height in an exponential way and it is halved each ~6500 m. On the contrary water pressure increases linearly with depth: 10 m of water increase the pressure of 1 atmosphere (760 mmHg) whereas more then 25 km of height are not yet sufficient to decrease the air pressure of one atmosphere (Fig. 3.3). This is because water has a density of 1000 kg/m^3 whereas air at sea level has a density of 1.275 kg/m^3. The rapid initial change of air pressure, and therefore of oxygen pressure, with altitude must be taken into account in a cardio patient together with his psychological effects due to the different surrounding, on the mountains versus a city.

3.4 Dynamics of Respiration

As blood flow in circulation, airflow in respiration is maintained by an intermittent action of muscles, the heart in circulation, the *respiratory muscles* in respiration. In circulation work has to be done to overcome the viscous friction of blood, in respiration work must be done to overcome viscous resistance of air and viscous resistance of lungs and chest wall. In circulation we have seen that the relationship between driving pressure ΔP and flow $\dot{V}$ of blood was defined by the law of Poiseuille for laminar motion: $\Delta P = (8l\eta/\pi R^4) \times \dot{V}$ (see Sect. 1.4.1) and by $\Delta P = a\dot{V}^x$ with $x > 1$ for turbulent motion (see Sect. 1.4.4). In respiration the equation used is Rohrer's:

$$\Delta P = a\dot{V} + b\dot{V}^2$$

where the first term refers to the laminar flow and the second term to the turbulent flow (a and b are two empirical constants).

As described in Sect. 1.4.4, the critical velocity is defined by the equation: $v_{crit} = N_c\eta/dr$. The viscosity of air at body temperature is $\eta_{air} \approx 0.02$ centipoise versus that of blood $\eta_{blood} \approx 4$ centipoise, i.e. η is ~200 times lower in air. As mentioned above, the density of air at sea level and body temperature is about 1 kg/m^3 whereas that of blood is 1000 kg/m^3, i.e. d is ~1000 times lower in air. It follows that the critical velocity $v_{crit} = N_c\eta/dr$ is 5 times lower in air than in blood. In the respiratory apparatus turbulence occurs preferably in the first airways (trachea and bronchi) than in the smaller airways near the alveoli because the radius r is greater in the first airways. Similarly to circulation (Fig. 1.18), the area of the overall section increases from trachea to bronchioles, and since the total flow is the same, the velocity in each conduct decreases from trachea to bronchioles. It follows that turbulence, and related noise, are more likely to occur at the level of the first airways. This must be taken into account when listening the lungs of a patient with a phonendoscope.

3.4.1   The Respiratory Muscles

Whereas in circulation the "motor" is represented by the two heart ventricles, in respiration it is represented by the respiratory muscles: inspiratory muscles and expiratory muscles. The *inspiratory* muscles are represented by the *diaphragm* and by the *external intercostal* muscles. The *expiratory* muscles are represented by the *abdominal* muscles (pushing up the diaphragm) and by the *internal intercostal* muscles. Contrary to stroke volume in circulation, which remains 'actively' constant with increasing muscular exercise, in respiration the volume of air inspired at each breathe increases with increasing muscular exercise. The most important inspiratory muscle is the diaphragm: at rest the diaphragm is responsible for the 75% of the volume of the inspired air, this percentage decreases with increasing lung ventilation.

As shown in Fig. 3.1 the input and output of air into e from the lungs is determined by the motion of the chest wall, mainly the diaphragm (the piston in Fig. 3.1). In Fig. 3.4 it is shown what really happens: the picture on the left shows the end of a forced expiration: it can be seen that the diaphragm and the intestinal mass are lifted squeezing air out of the lungs whose volume attains a minimum (darker area). The picture on the right shows the condition attained at the end of a maximal inspiration:

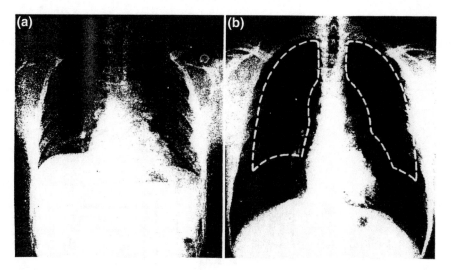

Fig. 3.4 Roentgenogram of the same chest in full expiration (**a**) and full inspiration (**b**). Dashed white line in (**b**) is outline of lungs in full expiration (as in (**a**)) (from Comroe JH Jr. *Physiology of Respiration*, 2nd ed., Year Book; 1975)

the darker area, representing the lungs volume, is increased well above the minimum (indicated by the interrupted lines) mainly thanks to an expansion of the basis of the lungs due to a lowering of the diaphragm.

Figure 3.5 shows that the intercostal muscles act as levers having the fulcrum near the vertebral column. The external intercostal muscles attach upward near the fulcrum and downwards to a distance from the fulcrum; it follows that when they shorten the lower point of insertion will be lifted cooperating to an increase in lung volume, i.e. the external intercostal muscles are inspiratory. The contrary is true for the internal intercostal muscles.

3.4.2 Start-End Points of Air Motion

Whereas in the unidirectional path of circulation the starting point of blood are the ventricles and the end point are the atria, in the bidirectional path of respiration the initial and the end points of air motion are the nose and the mouth to the alveoli in inspiration, and viceversa in the expiration. The air in the alveoli is surrounded by tissues also subjected to opposite motion during inspiration and expiration: the *lung* tissue, the *visceral* pleural, the *parietal* pleura and the *chest wall* represented by the *diaphragm* and the *rib cage*. A way to measure the pressure required to move the *air and the lungs* is by an *esophageal balloon*: i.e. a catheter surrounded in its final tract by a thin wall of rubber containing air and swallowed into the *third half of the esophagus* with its initial tract connected to a manometer in contact with the

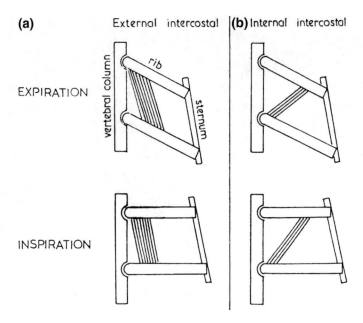

Fig. 3.5 Movements of the ribs induced by contraction of the external (**a**) and internal (**b**) intercostal muscles (from Comroe JH Jr. *Physiology of Respiration*, 2nd ed., Year Book; 1975)

atmosphere. A way to measure the pressure required to move the *air*, *the lungs and* the *chest wall* in a curarized subject is to place the subject in an *iron lung*. The dynamic pressure necessary to overcome the viscous resistance offered by the expansion of the lung tissue is greater than that offered by viscosity of air motion; at rest it represents 70% of the total; this percentage decreases with increasing ventilation, i.e. with the velocity of motion between air and tissue layers.

3.4.3 Effect of Air Motion and Lung Volume on the Airway Size

The airways diameter depends on two factors: (i) the difference in air pressure across their walls and (ii) the elastic traction on their walls by the surrounding pulmonary tissue (Fig. 3.1).

The first factor is an increasing function of the *velocity* of air motion and has an opposite effect during inspiration and expiration: during inspiration air pressure is lower outside (in the alveoli) than inside the *intra pulmonary airways*, and this difference in air pressure tends to expand the intrapulmonary airways; in contrast the *extra pulmonary* airways will tend to collapse during inspiration (an event prevented

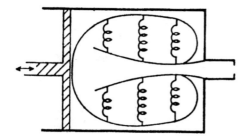

Fig. 3.6 As Fig. 3.1 to show that the elastic lung tissue surrounding the intrapulmonary airways pulls on them the more the greater the lung volume caused by expansion of the chest wall (modified from Fry DL and Hyatt RE *The American Journal of Medicine*, 29: 672–689,1960)

by cartilaginous rings) and expand during expiration. The dynamic effects described above during inspiration are opposite during the expiration.

The second factor is a function of the *volume* of the lungs: the greater the volume the greater is the elastic traction of pulmonary tissue on the airways with the consequence that their cross section and *conductance* (i.e. 1/*resistance*) increases (Fig. 3.6).

In Fig. 3.7a, the ordinate is the airways resistance expressed as the ratio between a pressure and a flow: $cmH_2O/(l/s)$. This is correct because we remember that ΔP = Resistance × Flow and, as a consequence, *Resistance = $\Delta P/Flow$*. The approximately hyperbolic relationship experimentally found between Resistance and lung volume, i.e. Resistance × Volume = constant, implies that 1/Resistance = *Conductance* = (1/constant) × Volume, i.e. airways conductance increases linearly with lung volume (Fig. 3.7b). This is a static effect.

Over-dilatation of the airways may be caused by relaxation of bronchial tone (caused e.g. by injection of atropine) or by increased tissue tension (e.g. pulmonary fibrosis or lung tissue resection). Partial bronchial obstruction may result in narrowing of the airways when the lungs are expanded, and complete closure when the lungs are deflated. Abnormal slackness of the lung tissues, as in emphysema, results in a partial closure of the airways when the lungs are inflated and complete closure when the lungs are deflated. This is particularly evident when air is expired quickly (as when trying to switch off a candle). During normal breathing an emphysematous patient may spontaneously keep his lips partially closed during expiration to keep a positive air pressure in his upstream airways. As shown in Fig. 3.7 expansion of the lungs causes a dilatation of the intrapulmonary airways due to the elastic structures of the lung parenchyma pulling on them; when lungs elastic structures are partly destroyed, as in the pulmonary emphysema, their pull on thorax wall will decrease and thorax will expand even at rest (barrel chest).

The *dead space*, as mentioned above, is the volume of air (~150 ml in adult human) where no exchanges can take place between inspired air and blood due to the thickness of its walls. It consists of extra pulmonary and intra pulmonary structures, too thick to allow gas exchanges, which have several useful physiological functions. First, it is

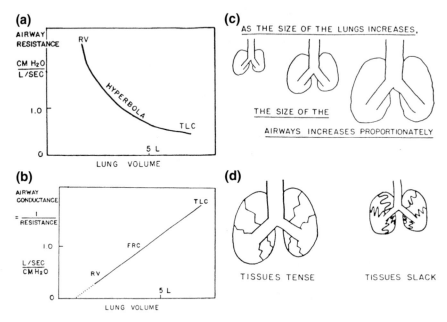

Fig. 3.7   a The airways resistance changes with lung volume (5 l = five liters) in an approximately hyperbolic relationship; RV: residual blood volume, TLC: total lung capacity (see *Static* section below). **b** Reciprocal of airways resistance, the *conductance*, increases linearly with lung volume. The extrapolation of the line to the *x*-axis intercept gives a value for the volume of the lung at which the airways close. FRC: functional residual capacity. **c** From the linear relationship plotted in (**b**), it is inferred that as the size of the lungs (diameter) increases the size of the airways (diameter and length) increases proportionately. **d** It is the traction of the lung tissues that keeps the airways open. *Left*: inflated position, tissues tense, airways wide open. *Right*: deflated position, tissue less tense, airways partly closed (modified from DuBois AB *Resistance to breathing* in Fenn WO and Rahn H (eds), *Handbook of Physiology* Section 3, Volume 1, American Physiological Society, Washington DC, 1964)

obvious that the delicate alveolar structure could not be exposed directly to contact with the external surrounding. Furthermore the inspired air in its transit through the wet and warm airways walls of the dead space reaches the alveoli saturated with water vapor and at body temperature (37 °C): this prevents a massive evaporation from the large surface alveolar area (~75 m^2), with a consequent large cooling of blood and dryness of the alveoli. In addition the curvatures of the air flow during inspiration, particularly in the nose and first airways, cause projection of smog particles on the wet dead space walls, where they are retained depurating the air reaching the alveoli.

3.5 Measurement of Lung Volumes

3.5.1 Spirometers

Spirometers are used to study the lung volume changes in different conditions (rest, exercise etc.). There are two kinds of spirometers: *closed circuit* (Benedict) and *open circuit* (Tissot).

Figure 3.8 shows the closed circuit Benedict spirometer. Unidirectional air flow is attained thanks to two valves consisting each in a thinner mobile strip pushed-pulled against a more rigid structure: the volume of air included between the two valves and the subject mouth is called *added dead space*. At the end of expiration this volume of air (say 20 ml) adds to the physiological dead space, 150 ml, with the consequence that the first 170 ml inspired at each breath are useless because they contain air with the same composition of the *alveolar air after* the exchange with blood. Aim of respiration is to set free of the CO_2 continuously produced by body's tissues and to replace the O_2 continuously absorbed. For this reason no single tube must be added between the breathing subject and the valves group of the spirometer (e.g. to cover the distance between subject bed and spirometer) because the subject would continue to inspire his alveolar air and die. In that particular condition *both tubes* must cover the distance between subject and spirometer thus keeping the dead space of the valves group near the subject mouth.

The amount of air breathed in or out during normal respiration is called *tidal volume*, V_t. The expired tidal volume $V_{te} = V_{ti} + y - x$ where V_{ti} is the inspired tidal volume, y is the volume of CO_2 added and x the volume of O_2 absorbed at each breath. Their ratio is called *respiratory quotient*: $QR = y/x = CO_2/O_2$. If the $QR = 1$ the *spirometric record* will be horizontal (as that in Fig. 3.8 right). In a system as that described in Fig. 3.8, the air re-respired in the spirometer will become richer in CO_2 and poorer in O_2. The first inconvenience can be eliminated by using a CO_2 *absorber* inserted in the expiration tube: a filter box containing substances capable to bind CO_2 (NaOH and $Ca(OH)_2$). In this case the volume of air in the spirometer will continuously decrease, the spirometer bell will lower, the spirometric record will progressively increase and the difference in its level will represent the volume of oxygen consumed V_{O_2} in the time interval considered, Δt (see Fig. 3.17). This method can be used to determine the oxygen consumption only at rest because the CO_2 absorber is able to absorb all the CO_2 produced only at low ventilation levels (*Ventilation* $= V_t \times f$, where f is the breath frequency). In addition one should take into account that the O_2 concentration in the spirometer will continuously decrease during the test. This method is used to determine the *basal metabolism*: $\dot{V}_{O_2} = V_{O_2}/\Delta t \approx 250$ ml/min in human at rest. As it will be described below, greater values of oxygen consumption are determined using the composition of expired air collected with the spirometer of Tissot (Fig. 3.9).

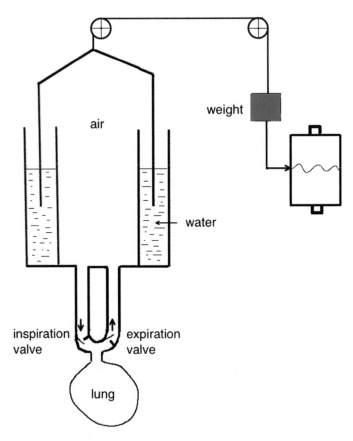

Fig. 3.8 Closed circuit Benedict spirometer. During inspiration, the inspiratory valve opens (left), the expiratory valve closes (right) and air is drawn from the spirometer bell into the lungs. During expiration, the inspiratory valve closes (left), the expiratory valve opens (right) and air flows from the lungs into the spirometer bell. A weight counterbalances the weight of the spirometer bell that is partly immersed in water. The spirometer bell movements during respiration are recorded on a rotating drum

3.5.2 *Divisions of Lung Volumes*

Figure 3.10 shows tracings of few breaths recorded by means of the Benedict closed circuit spirometer without an oxygen absorption device. We have already seen that the volume of air inspired and expired at each breath is called *tidal volume V_t* (V_{ti} during inspiration and V_{te} during expiration). If, at the end of normal expiration, the subject is invited to inhale air with a maximal inspiration, the total lung volume is attained; if this maximal inspiration is followed immediately by a maximal expiration, the volume of air expired corresponds to the maximal air volume the respiratory muscles can move: the *vital capacity*. The volume of air remaining in the lungs after a maximal expiration is called *residual volume* (Fig. 3.4a). *Actual values of these*

Fig. 3.9 Open circuit Tissot spirometer. Fresh air is inspired from the ambient and *expired* through the valve *t* into a bell partly immersed in water. When the bell is full of expired air, it is emptied through the three ways valve *t* by pushing it down by hand; expired air remains only in the tubes. *T* thermometer used to measure temperature of the expired air, *G* is a tap to withdraw expired air sample. The arrow attached to the weight, counterbalancing the weight of the bell, indicates a pen measuring volume changes on the scale *S*. *This spirometer is used to measure the volume and the composition of the expired air* (from Margaria R and De Caro L *Principi di Fisiologia Umana*, Francesco Vallardi, Milano, 1977)

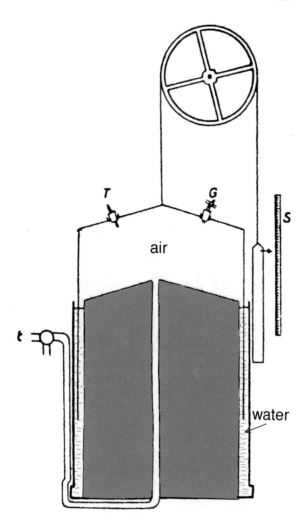

volumes vary considerably between subjects; 4–6 l for the vital capacity and ~1.5 l for the residual volume.

Note that the 'vital' capacity has no relation with the fitness of the subject or his capacity to perform muscular work; let's remember that the 'slower wagon' is the heart, not respiration. Note also that the *functional residual capacity* acts a *buffer*, which decreases the oscillations in CO_2 and O_2 concentrations in the alveolar air and in the blood. The two ways *Lungs Ventilation* = $V_t \times f$ (where $f = 1/(V_{ti} + V_{te}$ durations) can be compared with the one way *Cardiac output = Stroke volume* $\times f$ (where $f = 1/($systole + diastole durations); the important difference between the two is that with increasing muscular exercise *Lung Ventilation* increases thanks to an increase in *both* V_t and f, whereas cardiac output increases mainly thanks to an increase of f only (Fig. 2.49).

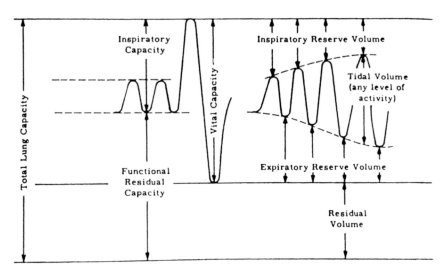

Fig. 3.10 Records obtained by means of a closed circuit spirometer without CO_2 absorber and with $QR = 1$ (Fig. 3.8). From left to right: the *total lung capacity* (total lung volume) is shown as the sum of the *inspiratory capacity* and *functional residual capacity* representing the volumes of air inside the lungs above and below a normal expiration. The *vital capacity* represents the maximum air the respiratory muscles can move in and out the lungs: the air remaining in the lungs after a maximal expiration is call *residual volume*. Above, right: lungs ventilation is increased by increasing the tidal volume inspiring a progressively greater air volume, utilizing the *inspiratory reserve volume*, and expiring a progressively greater air volume, utilizing the *expiratory reserve volume* (from Keele CA and Neil E *Samson Wright's applied physiology*, Oxford University Press, London, 1971)

The temperature of the air in the spirometer is usually lower than that in the lungs (37 °C). It follows that the volumes of air measured with the spirometer will be lower than those within the lungs and will change according room temperature. In order to get a measure of the air volumes within the body it is therefore *necessary to correct the volumes as measured by the spirometer* (Fig. 3.10). This is done using the *Boyle Law: PV = nRT*, where P and V are the pressure and the volume of gas, n is the number of moles, R is a constant and T is the absolute temperature. Since n and R are constant we can calculate the volume of air in the body V_b from that measured at ambient temperature V_a: $P_a V_a/T_a = P_b V_b/T_b$, i.e. $V_b = T_b/T_a \times P_a/P_b \times V_a$. Assuming an ambient temperature of 20 °C (i.e. 293 °K), $P_a = 740$ mmHg and subtracting the partial pressure of saturated water vapor at 20 °C (18 mmHg at 20 °C see Fig. 3.2):

$$V_b = (273 + 37)/(273 + 20) \times (740 - 18)/(740 - 47) \times V_a, \text{ i. e. } V_b = 1.1 V_a.$$

This correction is called ATPS to BTPS, i.e. Ambient, Temperature, Pressure, Saturated into Body, Temperature, Pressure, Saturated. For comparison purposes it is often necessary to express the pulmonary volumes in a standard condition instead of in the particular body condition described above. This correction is called the transformation ATPS into STPD, i.e. the transformation of respiratory volumes as measured in the ambient V_a into Standard Temperature Pressure and Dry volumes V_s, where standard temperature is $T_s = 273\ °K$ (i.e. 0 °C), and dry standard pressure $P_b = 760$ mmHg (i.e. one atmosphere without the pressure of the water vapor): $P_a V_a / T_a = P_s V_s / T_s$, i.e.: $V_s = T_s / T_a \times P_a / P_s \times V_a$, i.e., assuming $T_a = 20\ °C$:

$$V_s = (273/293) \times (740 - 18)/760 \times V_a = 0.9 V_a.$$

The maximum volume we can record directly with a closed circuit spirometer (Fig. 3.8) is the Vital Capacity, CV (Fig. 3.10). The *total* lung volume TLC = CV + RV where RV is the Residual Volume. Two methods are employed to measure the total lung volume TLC: (i) the *Helium Method* and (ii) the *Body Plethysmography Method*.

The *Helium Method* consists in the following procedure. A spirometer is washed several times with pure oxygen in such a way to be sure that oxygen is the only gas it contains. Subsequently a known volume of helium is added to the spirometer (this is easily done by looking the rise of spirometer bell when He is added to the circuit). Subsequently a subject (with its nose closed by a pincer) is invited to inspire ambient air up to *the maximum of its total lung capacity*, to expire within the spirometer and to breath attached to the spirometer long enough so that all the gases (nitrogen, helium, oxygen and carbon dioxide) are mixed up within his *lungs and the spirometer having, together, an unknown volume* V_x. Note that of the four gases present in the mixed lungs-spirometer volume the nitrogen was initially present in the lungs only. When the air in the lungs and in the spirometer are completely mixed up a sample of air is made from the spirometer bell to measure the fraction of helium F_{He} and another sample to measure that of nitrogen F_{N_2}. The volumes of helium and nitrogen are, respectively $V_{He} = V_x \times F_{He}$ and $V_{N_2} = V_x \times F_{N_2}$. Dividing the two equations one gets: $V_{He}/V_{N_2} = F_{He}/F_{N_2}$, i.e. $V_{N_2} = (V_{He}/F_{He}) \times F_{N_2}$. *If the respiratory quotient QR* = 1, the volume of nitrogen $V_{N_2} = (0.79 \times V_p)$ where V_p is the TLC (alveoli plus airways). It follows, from the above, that $V_p = (V_{He}/F_{He}) \times (F_{N_2}/0.79)$.

The *Body Plethysmography Method*: the subject is sitting in a closed *box* with his head outside the box and a tight, rigid collar around his neck (Fig. 3.11). Contrary to the spirometer the volume changes during respiration are measured *outside* the chest wall by means of a bell connected to the inside of the plethysmograph and isolated from the outside ambient by partial immersion in water (as the spirometers of Figs. 3.8 and 3.9). The subject sits on an air conditioner, which maintains constant the temperature of the air inside the plethysmograph. When the subject exerts a pressure ΔP measured on a mercury manometer connected to his mouth, the air inside his lungs is compressed, the volume of his lungs decreases and this decrease in volume ΔV is *recorded* by the plethysmograph bell.

In the *Boyle Law: PV = nRT*, we can consider the term *nRT* equal before and after compression of the air within the lungs (starting from the maximum lung volume TLC) because the increase in temperature ΔT due to the increase in pressure ΔP is prevented by contact with the large alveolar area kept at 37 °C by the flowing blood

Fig. 3.11 Schematic
representation of a
plethysmograph. The
continuous and the
interrupted lines (thin for the
lungs and thick for a bell
partially immersed in water)
indicate the condition before
respectively after the subject
exerts a positive pressure
against a mercury
manometer. When pressure
is exerted against the
manometer (nose closed),
the volume of the lungs is
reduced as indicated by a
lowering of the bell
(interrupted lines)

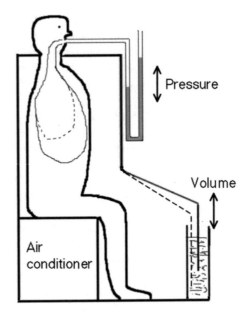

(T is kept constant during compression thanks to the rapid diffusion of heat from air
to the alveolar wall, see Diffusion in Sect. 3.9). It is therefore possible to write with
fair accuracy: $P_b V_p = (P_b + \Delta P)(V_p - \Delta V)$, from which the total lung volume V_p
(i.e. the TLC) can be measured since the barometric pressure P_b and all the other
terms of the equation are known. In the above equation we have neglected to subtract
from P_b the 47 mmHg due to the water vapor at 37 °C. The *total* lung volume TLC
$= CV + RV$ where RV is the Residual Volume: this can measured subtracting from
TLC, measured by the plethysmograph, the vital capacity CV measured by means
of a spirometer (Fig. 3.10).

What is left in our measurement of the different subdivisions of the lung volume
is the measure of the *dead space* DS. The volume of CO_2 expired equals the mixture
of that contained in the dead space and in the alveoli:

$V_{te} \times F_{eCO_2} = DS \times F_{iCO_2} + (V_{te} - DS) \times F_{alvCO_2}$. Since $F_{iCO_2} = 0.0003$
the first term of the equation $DS \times F_{iCO_2} \approx$ zero can be neglected. It follows that
$V_{te} \times F_{eCO_2} = V_{te} \times F_{alvCO_2} - DS \times F_{alvCO_2}$, i.e.

$$DS = V_{te} \times (F_{alvCO_2} - F_{eCO_2})/F_{alvCO_2} \text{ and, assuming } V_{te} = 500 \text{ ml,}$$
$$DS = 500 \text{ ml} \times (0.06 - 0.04)/0.06 = 500 \text{ ml}/3 = 160 \text{ ml.}$$

In order to measure F_{eCO_2}, and in general the *composition* and the volume of the
expired air in a given time, i.e. the *expiratory ventilation*, it is used the spirometer
of Tissot (Fig. 3.9). The subject inspires ambient air through the three ways valve
and expires in the Tissot spirometer (Fig. 3.12). When the bell of the spirometer is
completely lowered (by hands) a volume of air remains within its tubes and below

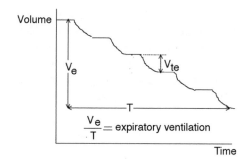

Fig. 3.12 Spirometric record obtained with the Tissot spirometer illustrated in Fig. 3.9. The horizontal tracts correspond to the time intervals during which the subject inspire from the ambient outside the spirometer. The average slope of the record represents the expiratory ventilation

the lowered bell. However if the procedure to expire in the spirometer and to lower the bell is repeated several times the composition of the volume of air expired into the Tissot spirometer will equal that coming from the lungs.

As described above, whereas in circulation there is a single unidirectional flow of blood (*cardiac output = Stroke volume $\times f$*), in respiration we have a two ways airflow: the *expiratory pulmonary ventilation* $\left(\dot{V}_{ep} = V_{te} \times f\right)$ and the *inspiratory pulmonary ventilation* $\left(\dot{V}_{ip} = V_{ti} \times f\right)$ where f is the respiratory frequency. $\dot{V}_e = \dot{V}_i$ when the respiratory quotient QR $= 1$. The *pulmonary inspiratory ventilation* $\dot{V}_i = V_{ti} \times f$ is only in part useful for gas exchanges because the first volume of air inspired is that contained in the *Dead Space, DS*, which contains the alveolar air of the previous expiration. The 'useful fresh' volume of inspired air is $V_{ti} - DS$ and the *alveolar ventilation is* $\dot{V}_{ia} = (V_{ti} - DS) \times f$. This shows that *the same pulmonary ventilation results in different alveolar ventilations depending on the tidal volume amplitude*; e.g.:

$$\dot{V}_{ip} = V_{ti} \times f = 500 \text{ ml} \times 20 \text{ min}^{-1} = 10 \text{ l/min};$$
$$\text{same as: } \dot{V}_{ip} = 1000 \text{ ml} \times 10 \text{ min}^{-1} = 10 \text{ l/min}$$
$$\dot{V}_{ia} = (V_{ti} - DS) \times f = (500 - 150)\text{ml} \times 20 \text{ min}^{-1} = 7 \text{ l/min, versus}$$
$$\dot{V}_{ia} = (V_{ti} - DS) \times f = (1000 - 150)\text{ml} \times 10 \text{ min}^{-1} = 8.5 \text{ l/min}$$

Disregarding the mechanical work done to sustain respiration, which will be treated later, deep breaths at a lower frequency will be more efficient than shorter breaths at a high frequency maintaining the same pulmonary ventilation: from this point of view the *respiratory efficiency* is greater the greater V_{ti} at a given $\dot{V}_{ip}$. An interesting phenomenon, which applies to the above description, is that of *panting*. Dogs and cats do not sweat and therefore, when the ambient is hot, they cannot lose heat by evaporation of water over their whole body skin. They lose heat by *panting*, i.e. by increasing $\dot{V}_{ip}$ maintaining the required $\dot{V}_{ia}$ thanks to a small $(V_{ti} - DS)$ and a high frequency f; in this way they increase water evaporation from the tongue and dead space favoring refreshing of the body while maintaining the alveolar ventilation appropriate for their metabolism; in other words only a small fraction of V_{ti} enters the alveoli whereas a larger fraction is used for heat dissipation. With this strategy they increase $\dot{V}_{ip}$ without falling into *hyperventilation*, which is defined as an *excess of alveolar ventilation* relative to the metabolic requirements; *hyperventilation* results in *hypocapnia* (excessive decline in partial CO_2 pressure).

3.5.3   Flowmeter

As described in Circulation, the flow of blood can be measured by means of a flowmeter (Fig. 1.16). Figure 3.13 shows the relationship between records obtained with a spirometer measuring the volume changes (Fig. 3.8) and with a much simpler apparatus, measuring the flow of air (Fig. 1.16). The interrupted lines in Fig. 3.13 show the relationship between volume V and flow dV/dT by joining the maxima and the minima of the two functions (see legend of Fig. 3.13). In mathematical terms the flow is the time derivative of the volume. Conversely it can be understood the reverse, i.e. that the area below the $dV/dT = f(t)$ curve, having dimension $dV/dT \times dT = dV$ has the dimensions of a volume. The sum of an infinite number of rectangles ($\int$) having ordinate dV/dT and abscissa dT (represented mathematically as $\int dV/dT$, i.e. the *integral* of the record below in Fig. 3.13), results in the $V = f(t)$ record above in Fig. 3.13, i.e. in the spirometric tracing.

It is therefore possible in practice to integrate electronically the signal obtained using the simple device such as that in Fig. 1.16 in order to measure simultaneously flow and spirometric record (this simpler, but indirect procedure however is subjected to a possibility of error due to the drift of the integrator outputs).

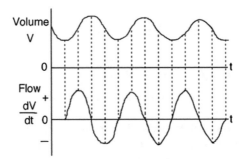

Fig. 3.13 Upper tracing: spirometric Volume (V) versus time (t) record obtained with a close circuit Benedict spirometer such as that in Fig. 3.8. Bottom tracing: flow metric record (dV/dt) derived from the slope of the upper tracing: the interrupted lines show when the flow (i.e. the slope of the upper tracing) is nil at the end of expiration and inspiration, maximal during inspiration (positive dV/dt records) and minimal during expiration (negative dV/dT records)

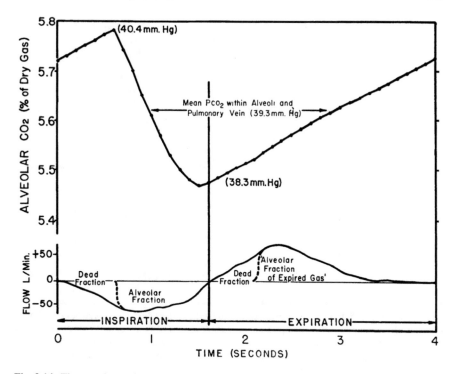

Fig. 3.14 The experimental tracings plotted above show the changes in alveolar CO_2 concentration and partial pressure during one respiratory cycle. The record below shows the simultaneous airflow in-out the lungs during inspiration and expiration. The areas between the flow record and the abscissa represent the total volume of air inspired and expired with the fraction contained in the dead space (from DuBois AB, Britt AG and Fenn WO *Journal of Applied Physiology* 4: 535–548, 1952)

Experimental records of the airflow into and out of the lungs are given in the bottom tracing of Fig. 3.14 (contrary to Fig. 3.13 the flow is represented negative during inspiration and positive during expiration). The area above and below the flow record represents the volume of air inspired and expired; the fraction of this volume corresponding to the dead space is indicated. The upper tracing in Fig. 3.14 shows the oscillation in CO_2 concentration in the alveolar air during one respiratory cycle. Note that the alveolar CO_2 *continues to increase* attaining a maximum at about the first third of the inspiration; this is because the first volume of air entering the alveoli is that of the dead space containing alveolar air of the previous expiration. During expiration the first volume of the air derives from the dead space, which did not undergo gas exchanges with blood. It can be seen that he minimum of CO_2 concentration is attained *before* the end of inspiration: this is because the input flow of fresh air at the end of inspiration is smaller then the *continuous* reversal flow of CO_2 from blood into the lungs. In conclusion the concentration of the *respiratory* gases CO_2 and O_2 continuously oscillate whereas the flow of CO_2 emission into the alveoli and O_2 absorption from alveoli into the blood are continuous processes.

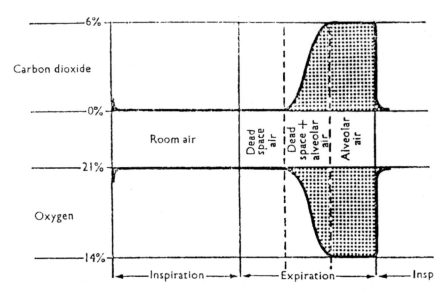

Fig. 3.15 Diagram showing the changes in composition of the respiratory gases recorded at the mouth (or nose) during a single respiratory cycle. During inspiration the composition is that of room air. During expiration the composition changes from that of room air (dead space air) to that of alveolar air at the end of expiration. An analysis of an end-expiratory sample will thus enable to determine the composition of alveolar air (from Green JH *An introduction to human physiology,* Oxford University Press, London, 1976)

Figure 3.15 shows records obtained by drawing airflow with a small tube connected to CO_2 and O_2 meters during inspiration and expiration. See table in the Introduction of this chapter where the composition of mixed expired air is also given. Instead of plateauing as simplified in Fig. 3.15 the concentration of CO_2 increases slightly and that of O_2 decreases slightly at the end of the expiration. A sample of alveolar air must be taken at the end of the expiration.

3.6   Metabolism

The physical dimensions of the metabolism are *work/time = power*. In the MKS system its units of measure are *Joules/sec = Watt*. However the units usually utilized to measure the metabolism is the *kilocalorie (kcal or Cal) in unit time* (remember that 1 kcal is the amount of energy necessary to increase the temperature of 1 l of water from 14.5 to 15.5 °C). Another unit of energy used is the *kilogrammeter = 2.34 small calories* (1 *small calorie* = 1/1000 Cal). We will consider the *basal metabolism* and the *energetic metabolism*. In a car the power is increased by increasing the gas burned in unit time, in our body the metabolism is increased by increasing burning in unit time of chemical energy of *glucides, proteins* and *lipids*. The chemical energy is transformed during the metabolism in others kinds of energy: heat, chemical energy (e.g. liver), osmotic energy (e.g. kidney), electrical energy (e.g. nervous system) and, particularly evident, into mechanical energy during muscular exercise. Note that whereas the brain metabolism remains practically the same during sleep and during intense cerebral activity, this does not mean that its energy expenditure is small: in a normal subject the brain represents about 2% of total body mass, but consumes about 20% of the total metabolism at rest.

3.6.1 Basal and Energetic Metabolism

The *basal metabolism* is the minimal power necessary to sustain the vital functions (e.g. cells turnover). It must be measured in the subject lying at a *comfortable* temperature, and fasting from at least 12 h to avoid the *specific dynamic action of food* (the amount of energy expenditure above the basal metabolic rate due to the cost of processing food for use and storage). The basal metabolism is expressed in kcal/(square meter of body surface and hour); it changes with age: in a 20 years old subject is 40 kcal/(m^2 × hour) whereas in a 70 years old subject is 35 kcal/(m^2 × hour). The body surface is ~1.75 m^2. In a day the difference is 5 kcal × 1.75 × 24 = 210 kcal. Running requires an energy expenditure of 1 kcal/(kg × km). Assuming a body mass of 70 kg, i.e. 70 kcal/km, the older subject should run 3 km everyday to maintain the body weight equal that of the young man!

As an example of *energetic metabolism* let's consider again running and a race of 20 km of a 70 kg subject: 1 kcal/(kg × km) × 70 kg × 20 km = 1400 kcal. Since burning a gram of fat evolves ~10 kcal, the total weight loss (excluding evaporation of water) will be only 140 g. This proves very useful when animals and humans must cover large distances without feeding. Another example: the energy expenditure to climb 100 m. In this case it is useful to use as energy unit the *kilogram meter,* *which corresponds to 2.34 small calories.* The *work* done against gravity will be 70 kg × 100 m × 2.34 = 16,380 cal. As we know the Efficiency = Work/Energy expenditure = 0.25, it follows that the energy expenditure will be Work/Efficiency, i.e. 16 kcal/0.25 = 64 kcal.

3.6.2 Methods of Measurement of Metabolism

Metabolism can be measured by *direct* and *indirect* methods. The direct method is illustrated in Fig. 3.16.

The chemical potential energy possessed by our organism is continuously transformed into work and heat in everyday life. If the work done is not stored as potential energy (for example lifting a weight from below to above a table), but is transformed into heat by frictions (e.g. in respiration and blood circulation), all the chemical energy is degraded into heat, which is a measure of the power, i.e. of the chemical

Fig. 3.16 Schematic representation of the Atwater-Rosa-Benedict respiration calorimeter to determine the heat produced from the difference in temperature from input and output of the water (gray) surrounding the chamber

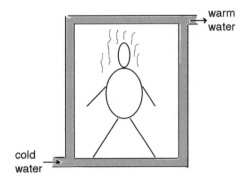

energy used in unit time to maintain life. In Fig. 3.16 is shown a method used to measure metabolism from the increase in temperature of water in equilibrium with subject's heat production (the increase of 1 °C of 1 l of water requires 1 kcal). This method is complex to realize because all the objects contained in the calorimetric room must attain the same body temperature otherwise they will absorb or emit heat instead of the body.

In order to determine the metabolism, i.e. the chemical energy used in unit time, we may in principle measure the amount of glucides, proteins and lipids intake, but this procedure would be practically impossible given the water content of the food ingested and the losses taking place with the excretions. Much more convenient is the use of *indirect* methods, based on the measure of the oxygen consumption once it is known its *caloric equivalent*, i.e. the energy set free when O_2 is used to burn proteins, lipids, glucides and a mixture of them. Consider for example the combustion of one mole of glucose: $C_6H_{12}O_6 + 6O_2 = 6CO_2 + 6H_2O + 670$ kcal.

Figure 3.17 shows one method to measure the O_2 consumed. As mentioned previously, this method is based on the complete absorption of the CO_2 produced. Since $V_{te} = V_{ti} + CO_2 - O_2$, if CO_2 is completed absorbed, $V_{te} < V_{ti}$ resulting in a drift of the spirometric record indicating the O_2 consumed (Fig. 3.17 right). This method however is adequate only for low values of ventilation ($V_{te} \times f$) due to the limit posed by the incomplete absorption of CO_2 when the velocity of air through the CO_2 absorber is too high.

During exercise the oxygen consumption is determined with the *open circuit* method (Fig. 3.9) based on the equation: oxygen consumed = oxygen inspired − oxygen expired, i.e. $\dot{V}_{O_2} = \dot{V}_i F_{iO_2} - \dot{V}_e F_{eO_2}$. The only term unknown in this equation is $\dot{V}_i$: in fact $F_{iO_2} = 0.2093$, $\dot{V}_e$ equals the average slope in Fig. 3.12 and F_{eO_2} can be measured by analyzing the expired air in the Tissot spirometer. In order to know the only unknown term, $\dot{V}_i$, consider that the volume of nitrogen inspired in unit time equals that expired because nitrogen is an inert gas, which does not participate to respiratory exchanges: $\dot{V}_i, F_{iN_2} = \dot{V}_e F_{eN_2}$, i.e. $\dot{V}_i = \dot{V}_e F_{eN_2}/F_{iN_2}$. The fraction of nitrogen inspired is $F_{iN_2} = 0.7903$ whereas he fraction expired F_{eN_2} can be determined by analyzing the composition of the expired air by withdrawing a sample of the air exhaled in the Tissot spirometric bell as described in Fig. 3.9. Note that $\dot{V}_i = \dot{V}_e$ if

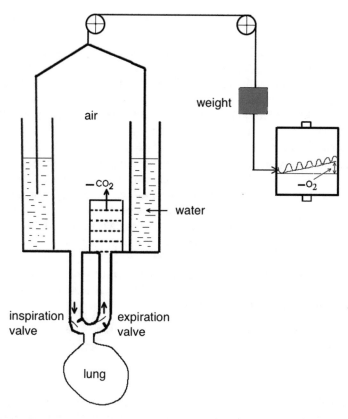

Fig. 3.17 Closed circuit Benedict spirometer as that described in Fig. 3.8 with an added box in the expiratory line with chemicals capable to absorb the carbon dioxide CO_2 produced by the breathing subject. The lowering of the spirometer bell with time (tracing on the right) allows determining the volume of oxygen O_2 consumed. Other indications as in Fig. 3.8

the respiratory quotient $QR = CO_2/O_2 = 1$. It follows that the oxygen consumption equals: $\dot{V}_{O_2} = (\dot{V}_e F_{eN_2}/F_{iN_2})F_{iO_2} - \dot{V}_e F_{eO_2}$.

In order to determine the metabolism, once known the volume of oxygen consumed, it is necessary to know the *caloric equivalent of the oxygen consumed*, i.e. the *kcal/liter of oxygen consumed*. In principle, oxygen can be used to burn glucides, lipids and proteins. Therefore, in order to know the caloric equivalent of the oxygen consumed, we must know how much of it is used to burn glucides, fats and proteins and the heat produced by burning one gram of each of them. Of these three constituents, mainly the glucides and the lipids represent the fuel burned to produce work and heat, whereas the proteins are mainly used to construct, repair and substitute tissues. We can therefore fairly well solve this problem by neglecting the heat produced by burning proteins, and by measuring the fraction lipids/glucides burned from the *non proteic respiratory quotient*: $NPRQ = 0.7$ when burning lipids only,

$NPRQ = 1$ when burning glucides only and intermediate when burning a mixture of the two.

Since it is known that 1 l of oxygen corresponds to 4.7 kcal when burning fat and 5 kcal when burning glucides, it is possible to determine the caloric equivalent of oxygen consumed over the whole $NPRQ = 0.7$–1 range (see Table 3.4). We can calculate the $x\%$ of lipids burning as follows: $(1 - 0.7): 100\% = (1 - NPRQ): x\%$, i.e. $x\%$ lipids $= 100(1 - NPRQ)/(1 - 0.7)$. Usually $NPRQ = 0.825$ at rest corresponding to a caloric equivalent of oxygen consumed of 4.8 kcal/l.

It must be pointed out that the *kcal/liter of oxygen consumed* corresponds to the total metabolic energy output only in *aerobic* conditions, i.e. in exercise (e.g. running) that can be maintained endlessly without incurring into an *oxygen debt* (production of lactic acid, H^+L^-). In case lactic acid is produced the $NPRQ$ increases because additional CO_2 is set free from the reserve of bicarbonate according to the following equation: $N_a{}^+ + HCO_3{}^- + H^+ + L^- = N_a{}^+ + L^- + H_2CO_3 = N_a{}^+ + L^- + H_2O + CO_2$. Another factor, which may affect $NPRQ$, is the transformation of one element into another. For example in a diet rich in glucides (e.g. $C_6H_{12}O_6$) the transformation of glucides into the fat molecule, poorer in O_2, causes a reduction of $NPRQ = CO_2/O_2$ measured with the spirometer because some O_2 is set free internally instead of being absorbed during respiration.

3.6.3 The Oxygen Debt

Figure 3.18 shows that the oxygen consumed in unit time by the human 'motor', contrary to that of a car, does not increase immediately from zero to the value required by the imposed load and does not fall immediately to zero when unloaded. In human, oxygen increases from the metabolism at rest to the value required by the imposed load at first rapidly and then more slowly. Similarly, at the end of the imposed work, oxygen consumption does not fall immediately, but decreases at first rapidly and then more slowly towards the resting value. The areas between the interrupted and the continuous line in Fig. 3.18 ($\dot{V}_{O_2} \times \Delta t$) represent a *volume of oxygen*: the *acquisition of the oxygen debt* at the beginning of the exercise and the *payment of the oxygen debt* after the end of the exercise. The difference between interrupted and continuous lines is due to the fact that whereas in the car the motor burns immediately gasoline using oxygen to produce work, in human, as we have seen (Chap. 2), the first and final immediate energy source is ATP. The oxygen consumption follows work production at the beginning of the exercise and continues after the end of it to resynthesize the ATP immediately split during work production. Mechanical power is developed instantaneously (interrupted line), but time is required to resynthesize, by oxidative processes, creatine phosphate (CP) from creatinine, and ATP from ADP. The continuous and the interrupted lines join when splitting equals resynthesis of *phosphagen* (CP + ATP).

When the work imposed in unit time exceeds the *maximal aerobic power* and *lasts for a sufficient long time*, an additional power output contributes together with oxygen

Table 3.4 From Margaria R and De Caro L *Principi di Fisiologia Umana*, Francesco Vallardi, Milano, 1977

NPRQ	% heat produced by burning		1 l of oxygen corresponds to		
	Carbs	Fat	kcal	g carbs burned	g fat burned
0.707	0	100.0	4.686	0	0.502
0.71	1.1	98.9	4.690	0.016	0.497
0.72	4.8	95.2	4.702	0.055	0.482
0.73	8.4	91.6	4.714	0.094	0.465
0.74	12.0	88.0	4.727	0.134	0.450
0.75	15.6	84.4	4.739	0.173	0.433
0.76	19.2	80.8	4.751	0.213	0.417
0.77	22.8	77.2	4.764	0.254	0.400
0.78	26.3	73.7	4.776	0.294	0.384
0.79	29.9	70.1	4.788	0.334	0.368
0.80	33.4	66.6	4.801	0.375	0.350
0.81	36.9	63.1	4.813	0.415	0.334
0.82	40.3	59.7	4.825	0.456	0.317
0.83	43.8	56.2	4.838	0.498	0.301
0.84	47.2	52.8	4.850	0.539	0.284
0.85	50.7	49.3	4.862	0.580	0.267
0.86	54.1	45.9	4.875	0.622	0.249
0.87	57.5	42.5	4.887	0.666	0.232
0.88	60.8	39.2	4.899	0.708	0.215
0.89	64.2	35.8	4.911	0.741	0.197
0.90	67.5	32.5	4.924	0.793	0.180
0.91	70.8	29.2	4.936	0.836	0.162
0.92	74.1	25.9	4.948	0.878	0.145
0.93	77.4	22.6	4.961	0.922	0.127
0.94	80.7	19.3	4.973	0.966	0.109
0.95	84.0	16.0	4.985	1.010	0.091
0.96	87.2	12.8	4.998	1.053	0.073
0.97	90.4	9.6	5.010	1.098	0.055
0.98	93.6	6.4	5.022	1.142	0.036
0.99	96.8	3.2	5.035	1.185	0.018
1.00	100.0	0	5.047	1.223	0

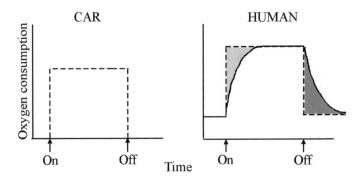

Fig. 3.18 The record on the left shows the oxygen consumption in unit time of a car: starting from zero, when the car is at rest, increases immediately to a given value as the motor is running and decreases immediately to zero when the motor is turned off. In contrast, the oxygen consumption during human exercise (continuous line on the right): (i) starts from the resting metabolism and (ii) attains its maximum and falls to zero with a progressively slower trend. The gray areas represent the *volume* of oxygen consumed in defect (*acquisition of oxygen debt*, lighter gray) and in excess (*payment of oxygen debt*, darker gray) to that required by the imposed load (interrupted line)

consumption to resynthesize phosphagen: the anaerobic glycolysis of glycogen, i.e. the transformation of glycogen into *lactic acid*. In this case the payment of the oxygen debt (darker gray area in Fig. 3.18) is greater than the acquisition of oxygen debt (lighter gray area in Fig. 3.18) because at the end of the exercise additional oxygen is consumed to revert lactic acid into glycogen. In a jump or in the 100 m race the power is greater than the maximal aerobic power and an oxygen debt is acquired, but no lactic acid is produced. This is because the exercise duration is too short to deplete the phosphagen energy sources (ATP + CP).

The payment of oxygen debt after exercise (darker gray area in Fig. 3.18) may therefore be made up by several components: (i) the obligatory oxygen consumed to pay the acquisition of the oxygen debt: the *alactacid debt* which is always present and has a half reaction time of 30 s, (ii) the eventual *lactacid debt* to convert lactic acid back to glycogen which has a half reaction time of 15 min (the 'day after' muscle pain therefore cannot be attributed to lactic acid, but to lesion of muscle fibers: in fact it is more frequent when the exercise requires the high forces attained during muscle stretching, e.g. running vs. cycling); (iii) a slight increment of the oxygen consumption at rest (Fig. 3.19).

3.7 Statics of the Respiratory System

As in the statics of circulation, the statics of the respiratory system studies the pressure changes taking place in absence of flow. In the statics of circulation the blood pressure changes are caused by gravity. Even if gravity does change the shape of the respiratory apparatus (e.g. standing vs. laying), the pressure changes of the air within the lungs

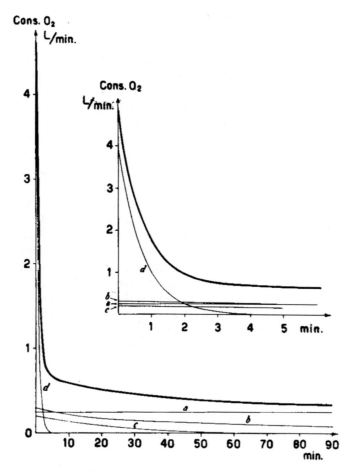

Fig. 3.19 Oxygen consumption in liter/minute after the end of muscular work (thick line). Thinner line curves correspond to the single elementary processes composing the thick line: (a) oxygen consumption corresponding to the basal metabolism, which is constant; (b) increment of the resting oxygen consumption tending to zero; (c) oxygen consumption corresponding to payment of the lactacid debt, (d) oxygen consumption corresponding to payment of the alactacid debt. In the center of the Figure is reproduced the initial part of the same graph to make more evident the trend of the faster processes (from Margaria R and De Caro L *Principi di fisiologia umana*, Vallardi, Milano, 1977)

are determined mainly by the tension (passive and/or active) of the walls surrounding the alveoli. The active tension is due to the contraction of the respiratory muscles. The passive tension depends from the volume and is represented by the *elastic tension* of the tissues and the *surface tension* at the interface between liquid on the alveolar surface and alveolar air.

3.7.1 Pressure-Volume Diagram of Living Lung

Figure 3.20 shows two different procedures to measure the pressure-volume diagram of the lungs. The procedure more commonly used (left) is to measure the volume by means of a syringe pushing and withdrawing air from the isolated lung. This procedure is subjected to an error due to the fact that some air is trapped into the lung due to airways closure (interrupted line). The continuous line in Fig. 3.20 shows the correct pressure-volume diagram obtained by measuring the volume changes of the open-chest living animal laying into a plethysmograph. In this case, the volume of lungs filled with oxygen decreases down to zero (complete atelectasis) because gas is absorbed from the blood independently of airways closure.

When the thorax is open (as in the schema of Fig. 3.20) the tendency to collapse of the lung is due only to its *weight*, to the *elastic tension* of its walls and to the *surface tension* at the interface between alveolar air and liquid on the alveolar surface. The *minimal air* volume is that contained in the lungs when their PV diagram crosses the ordinate, i.e. when the pressure inside the lungs equals that outside the lungs. At this volume the lungs surface is dark red in contrast with the light pink color at their maximal expansion. The minimal volume of the lungs in situ, the *residual air*, is always greater than the minimal air, i.e. the lungs in situ always tend to collapse. If absorption of oxygen is allowed to continue beyond this volume, the pressure across the lung becomes negative, i.e. lower than atmospheric, and the color of the lungs approaches that of liver. *The alveoli resist collapse until they become atelectatic, but do not tend to re-expand.* Expansion can be obtained by pushing air into them by

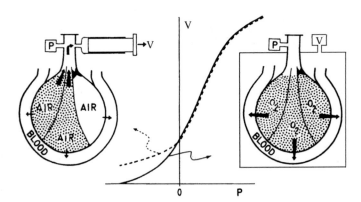

Fig. 3.20 Schematic representation of two experimental procedures to measure the pressure-volume diagram of the lungs. Left: the volume changes of the isolated lung filled with air are measured by means of a syringe; the interrupted line on the center of the Figure indicates the pressure-volume diagram so obtained. Right: volume changes of the lungs filled with oxygen are measured by means of a plethysmograph where the open-chest *living* animal was placed. The lungs are filled with oxygen and the volume is reduced to zero by the oxygen absorption by blood as indicated by the full line pressure-volume diagram (modified from Cavagna GA, Stemmler EJ & DuBois AB, *J. App. Physiol.* 22: 441–452, 1967)

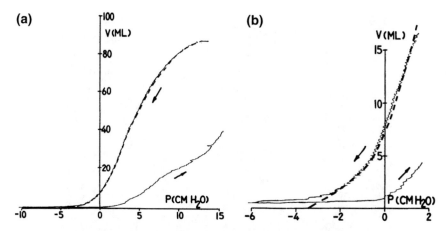

Fig. 3.21 a Pressure volume diagrams of lungs of a rabbit with the chest open as obtained during slow air withdrawal from the trachea (thinner continuous line) and during oxygen absorption by the blood (thicker interrupted line). **b** Only the lower part of the PV diagram was recorded for greater detail. Note that the continuous line deviates from the broken and becomes asymptotic to the abscissa before zero volume is reached. At this point movement of the syringe was reversed and PV diagram during inflation was partially recorded (from Cavagna GA, Stemmler EJ & DuBois AB, *J. App. Physiol.* 22: 441–452, 1967)

means of a syringe. The *opening pressure* of the alveoli is not equal for all of them, with the consequence that, during re-expansion it is possible to see atelectatic areas (dark) adjacent with over-expanded areas (white-pink). Re-expansion of atelectatic areas causes the indents observable on the PV diagram (Fig. 3.21a). Deflation of lungs is uniform whereas its re-expansion is irregular. Partial atelectatic areas can occur also in situ in some conditions (breathing oxygen, laying posture) due to the closure of some small airways. Note that, in spite of the appearance, the alveoli are structures *in series* not in parallel; in fact, the pressure is equal in all of them whereas their volumes add up. It follows that *at a given lung volume the number of open alveoli is greater during deflation (each exerting a lower pressure) than during expansion (each exerting an higher pressure)*. The deflation of the lung is more uniform than its expansion. Since the alveoli are in series, the whole lung tends to collapse with a pressure equal that of each single alveolus.

The slope of the curves in Fig. 3.21, dV/dP, represents the *extensibility* or *compliance* of the lungs. The work done $\int P\mathrm{d}V$ during decrease and increase of the lung volume is represented by the area between the *ordinate* and the tracings indicated by the arrows in Fig. 3.21. The difference between the two areas is called *pulmonary hysteresis* and is smaller the greater the volume from which the lungs are re-expanded. Pulmonary hysteresis is due to: (i) the *elastic* hysteresis, i.e. the difference between mechanical energy stored and released by any elastic structure (Fig. 1.20b), this is not relevant in the lungs; (ii) the *opening pressure* required to expand atelectatic structures; (iii) the *different number* of alveoli open (greater during relaxation and smaller during expansion); (iv) the *hysteresis of the surfactant* film on the alveolar

Fig. 3.22 Insect 'walking' on the air-water interface. Note the inflexions of the liquid surface supporting the weight of the insect

surface. The last three mechanisms mentioned above depend on the *surface tension* at the air-liquid interface (the presence of liquid on the alveolar surface is proved by the fact that the expired air is saturated with water vapor). What is the origin of the surface tension?

Figure 3.23 explains that. On the abscissa is given the distance r among water molecules. Different kinds of forces act between molecules and the resultant of these forces is indicated by the continuous line as a function of the distance between their center of mass. Above a given distance r_o the molecules attract each other: attraction increases up to a given distance beyond which it decreases. Below r_o the molecules repel each other. At the distance r_o equilibrium is reached between attraction and repulsion and the potential energy between them is at a minimum: this is the distance between molecules in the bulk of the liquid. The molecules at the liquid-air surface have no repulsion from water molecules on one side: therefore the distance between them increases until equilibrium is reached from the forces of attraction of water molecules and their tendency to evaporate. As a consequence of the greater distance between molecules on the surface relative to the equilibrium position a net force of attraction acts among them. *This force parallel to the surface is called surface tension* (Sect. 1.6.1) (Fig. 3.22).

The surface tension γ between the wet alveoli surface and the alveolar air tends to collapse the alveoli. Assuming a spherical shape of the alveoli ($\approx$700 millions of them for a total surface of $\approx$70 m^2, each with radius, $R \approx 50$–100 μm), each of them (and therefore their total assembly since the alveoli are in series) will tend to collapse with a pressure $P = 2\,\gamma/R$ (according to the law of Laplace, see Sect. 1.6.1). *The surface tension at the interface water-air is $\gamma = 72$ dyne/cm.* It follows that due to the surface tension water-air only the alveoli, and therefore the lung, should collapse, using the CGS system, with a pressure $P = 2 \times 72/100 \times 10^{-4} = 14{,}400$

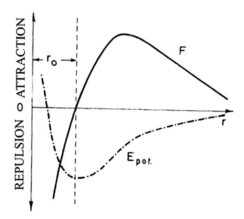

Fig. 3.23 The resultant of the forces of attraction and repulsion between molecules of a liquid F is plotted as a function of their distance r. The interrupted line indicates the potential energy E_{pot} due to their relative position, which attains a minimum at the distance of equilibrium r_0, i.e. when the forces of repulsion equal those of attraction (modified from Bernardini G, *Fisica sperimentale*, R. Pioda, R, 1940)

dyne/cm^2 $\approx$ 14 cm H$_2$O (30 cm H$_2$O if $R = 50$ μm). In addition, we should add the *elastic* tension of the structure of the alveolus around the water-air surface (like two rubber balloons one within the other). If we compare these figures with those on the abscissa of Fig. 3.21 during lungs deflation, when all the alveoli are open, we see that the lungs tend to collapse with a pressure less than that calculated above based on the water-air interface only.

This finding was first explained by Pattle RE (*Physiological Reviews*, 45: 48–79, 1965) studying patients affected by pulmonary edema. He noticed that the foam on the surface of glasses containing their sputum remained stable overnight. If we shake a water-detersive solution we get foam, but this disappears overnight because the potential energy at the surface tension-area interface of the bubbles tends spontaneously to a minimum. On the basis of this observation, Dr. Pattle concluded that a substance must exist in the lungs capable to reduce the surface tension to zero. This substance, which we will call *surface-active material*, is made up by bipolar molecules having an extremity hydrophilic and the other hydrophobic. If we 'wash' with a catheter a lung lobe and then withdraw the liquid, we obtain a solution containing surface-active material. In presence of a water-air surface molecules of surface-active material will be 'captured' on the surface with their hydrophilic extremity directed inside towards the bulk of the water mass and the hydrophobic extremity at the interface water-air. The ensemble of surfactant molecules on the surface forms a *film*, which tends to expand because surfactant molecules repel each other, opposite to the water molecules at the air-water interface, which attract each other as described above (Fig. 3.22). As known, the energy of a gas occupying a volume V with a pressure P is $P \times V = n\,R\,T$ (where n is the number of moles, R is a constant and T is the absolute temperature); similarly, the energy of a surfactant film occupying the surface A with a tension (force/length) π is $\pi \times (A - A_o) = n R T$ where A_o is the solid area occupied by the surfactant molecules packaged against each other. In other words the film tends to expand contrary to the surface tension at the pure water-air interface γ, which tends to collapse, with the result that the resultant surface tension at the film-water mixture will be $\gamma_S = \gamma - \pi$.

In addition to decrease the tendency of the alveoli to collapse, the surfactant has the important function to *stabilize* the alveoli. In fact, γ alone would tend to collapse the alveoli with $P = 2\gamma/R$, according to the law of Laplace, i.e. with a pressure P greater the smaller their radius. It follows that if an alveolus has a radius slightly smaller than that of the adjacent one, air will flow from the smaller to the larger alveolus more and more quickly with a positive feedback. This is prevented by the surfactant film because as soon as the radius and surface area A of the smaller alveolus tend to decrease, π will increase decreasing the surface tension γ_s and the tendency of the alveolus to collapse. The contrary is true in the adjacent alveolus: as soon as its surface area A tends to increase, π will decrease increasing the surface tension γ_s and the tendency of the alveolus to resist collapse.

The importance of surface tension on lung mechanics is evidenced in Fig. 3.24 where pressure-volume diagrams are made by filling and withdrawing liquid and air. It can be seen that: (i) the lung filled with liquid, where surface tension at the liquid-air interface is eliminated, tends to collapse with a pressure much smaller than the lung filled with air; this shows that *elastic recoil* alone has little effect on the tendency of the lungs to collapse; (ii) the *area of hysteresis* between stretch and recoil is practically nil in liquid filled lungs whereas hysteresis and the *opening pressure* are large in the air filled lungs, indicating that surface tension has a major impact on lung mechanics. The structure of the alveolus is like that of two balloons one within the other: an elastic tissue covered by the film of the tension active material. The mechanical characteristics of the film alone have been studied as described in Figs. 3.25 and 3.26.

A very thin glass slide of width l is partially immersed in the liquid withdrawn from a lung lobe and connected to a force transducer through a very thin glass tube (Fig. 3.25). The liquid crawls on both sides of the slide and, by tending to reduce its total liquid-air surface to a minimum, pulls on each face of the slide with a force $F = \gamma_s l$ (remember that γ has the dimensions of force/length). Note that the slide is completely wet by the liquid, which make a continuous surface up and around it.

The apparatus described in Fig. 3.26 has been used to simulate the effect on lung mechanics of the changes in the area occupied by the surfactant during respiration. The liquid extracted by washing the lung was first left to stay for a while in a container to allow the molecules of the surface-active material to reach the liquid-air surface (*aging* of the film). Subsequently the area of the film was compressed and dilated by means of a barrier while the surface tension γ_s and the area occupied by the film were measured (Fig. 3.27).

In Fig. 3.27 the area is expressed as a percentage of the area occupied initially by the film (after aging) before the barrier starts to compress it. When the area is reduced the molecules adsorbed on the surface approach each other exerting a progressively greater surface pressure π and consequently a reduction of $\gamma_s = \gamma - \pi$ down to

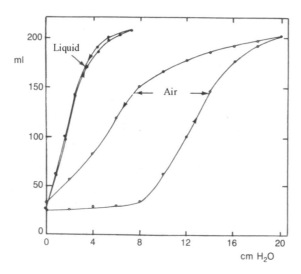

Fig. 3.24 Volume (ordinate)-pressure (abscissa) of a cat lung made from the *minimal air* up filling and withdrawing air and liquid (modified from Radford EP in *Tissue Elasticity*, Am. Physiol. Soc. Washington DC, 1957)

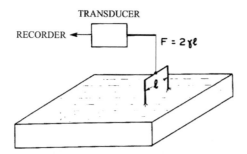

Fig. 3.25 Set up used to measure the surface tension γ_s of the liquid coming from inside the lung by measuring the force F with which each surface of the slide is pulled down by the tendency of the surfactant film to retract (modified from Hildebrandt J and Young AC. In Ruch TC and Patton HD (eds.) *Physiology and Biophysics* W.B. Saunders Company, 1966)

values of 0–3 dyne/cm. Evidently the maximal surface pressure of the compact film approaches the surface tension of water-air interface γ. If after reaching the minimum value of γ_s, the barrier allows the film to expand, γ_s increases first abruptly up to about 30 dyne/cm. a value similar to that of plasma suggesting a rupture of the film. If the movement of the barrier is stopped, one observes a progressive increment of γ_s during compression, as if molecules would leave the surface for the liquid phase, and vice versa during expansion (thick horizontal lines). Connecting the values of γ_s attained at equilibrium one obtains an area of static hysteresis (interrupted line) smaller than the dynamics one (thinner external line). The findings described in Fig. 3.27 are useful to understand: (i) genesis of pulmonary atelectasis particularly

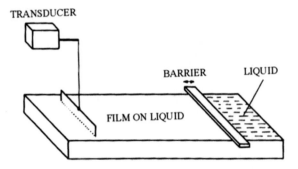

Fig. 3.26 Set up used to measure the area-surface tension diagram of the film covering the alveoli (from Hildebrandt J and Young AC. In Ruch TC and Patton HD (eds.) *Physiology and Biophysics* W.B. Saunders Company, 1966)

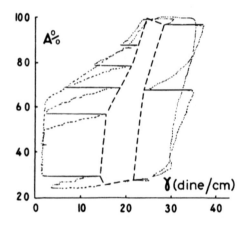

Fig. 3.27 Area-surface tension diagram obtained by means of the apparatus described in Fig. 3.26. Thinner lines were obtained during movement of the barrier. Horizontal thicker continuous lines indicate the change in surface tension when the movement of the barrier is stopped and surface tension reverts towards its state of equilibrium around 20 dyne/cm attained after aging of the film (interrupted line) (from Cavagna GA, Velasquez BJ, Wetton R and DuBois AB *J. Appl. Physiol.* 22: 982–989, 1967)

during assisted respiration and (ii) the necessity of the intermittent deep breathing. It is in fact possible that the small variation of the alveolar air surface as during normal and artificial breathing may lead to an increase of surface tension to relatively high values (~20 dyne/cm, Fig. 3.27) favoring atelectasis. It follows the necessity (to keep in mind during artificial respiration, particularly in anesthesia) to alternate deep inspirations, which allow opening small atelectatic zones and retrieving molecules of tension active material to the alveolar surface.

Some premature babies are born with a *hyaline membrane disease* due to the absence of pulmonary surfactant material. As described above their lungs will tend to collapse with a pressure greater than normal due to the high surface tension of the

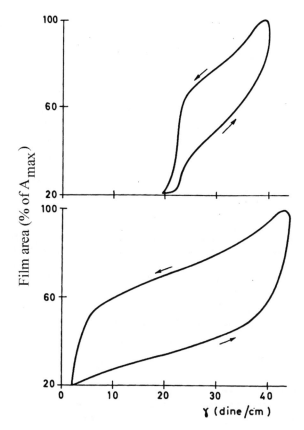

Fig. 3.28 Area-surface tension diagrams of the liquid extracted from the lungs of a baby dead because of hyaline membrane disease (above) and of a baby dead for other illness (below) (modified from Clements JA and Tierney DF in Fenn WO and Rahn H (eds.) *Handbook of physiology*, Sect. 3, vol. II, American Physiological Society, Washington DC 1965)

liquid wetting the alveoli γ not opposed by the surface pressure π (remember that $\gamma_s = \gamma - \pi$). In order to oppose the tendency of lungs to collapse, these babies are ventilated with a positive air pressure maintaining their lungs expanded. However, this procedure is not always successful; in fact the tension γ, unopposed by π, will cause a negative pressure between liquid-air surface and blood capillaries causing leak of plasma within the alveolus, i.e. a pulmonary edema. The more liquid enters the alveoli from the capillaries, the smaller will be the radius of curvature r of the liquid-air surface and the greater the pressure $P = 2 \, \gamma/r$ with which liquid is withdrawn from the capillaries (a positive feedback).

Figure 3.28 shows surface tension (γ)-area (% of maximal value before compression) diagrams made on lung tissue extracts from a baby dead for hyaline membrane disease (above) and for other cause (below). It can be seen that the area of hysteresis is smaller and the minimum value attained by surface tension is greater in the hyaline

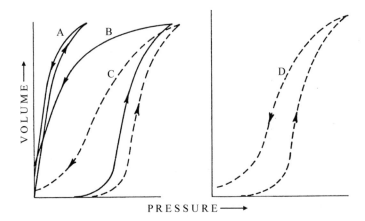

Fig. 3.29 Schematic representation of pressure-volume diagrams of lungs in different conditions. **a** Liquid filled. **b** Normal air-filled. **c** Air-filled hyaline-membrane. **d** Air-filled after detergent wash (modified from Hildebrandt J and Young AC *Anatomy and Physics of Respiration* in Ruch TC and Patton HD (eds) *Physiology and Biophysics*, WB Saunders Company, 1966)

membrane disease extract. Note that the maximum value of γ attained when the film is expanded is similar in both cases because of a reduction of surface pressure π takes place also in presence of surface-active material. However, when the volume (and the surface) are reduced after expansion, γ decreases to low values due to an increase of π in presence of surface active material, whereas γ maintains high values (about 20 dine/cm) up to the end of expiration in absence of surface active material.

A comparison of pressure-volume diagrams of the lungs obtained in the different conditions described above is schematized in Fig. 3.29. From left to right: liquid filled lung (A) showing the minimal hysteresis and tendency to collapse of the elastic structures only in absence of a liquid air interface. The normal air filled lung (B), differing from that affected by the hyaline membrane disease (C) mainly during deflation where pressure is much higher in the hyaline membrane lung due to the lack of surface pressure π; the trend during expansion is similar in both cases. Finally (D), the pressure volume diagram of a lung washed with a detergent, which does show hysteresis, but is unable to lower the pressure during deflation as the surfactant of a normal lung.

3.7.2 Intrathoracic and Intrapleural Pressure

Intrathoracic pressure equals the pressure in the mediastinum and in the esophagus (measured with an esophageal balloon) and differs from the intrapleural pressure, i.e. from the pressure in the thin space between visceral and parietal pleurae. Let's consider the intrathoracic, mediastinal pressure first.

Fig. 3.30 Schematic representation of the mechanism by which the elastic recoil of the lungs causes the intrathoracic depression of Donders

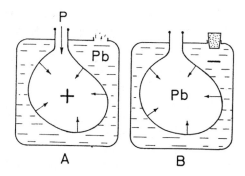

In Fig. 3.30 A: an elastic balloon immersed in a rigid vase full of water with a hole on its top is inflated by a pump to the pressure P while water spills out trough the hole. Neglecting the hydrostatic pressure due to gravity, the pressure of water equals the barometric pressure P_b because it is in communication with the outside through the hole. Inside the balloon the pressure P is positive (greater than atmospheric P_b) due to its tendency to collapse, as we have studied until now on the isolated lung (e.g. Fig. 3.29). B: now let's close the hole of the vase with a cork leaving the balloon open towards the ambient with barometric pressure P_b within it. The elastic walls of the balloon tend to collapse, but to do so the volume of water around them should increase, which is not possible because the water is inextensible. As a result the water is subjected to a negative pressure due to the elastic pull of the walls of the balloon. This is the origin of the *negative pressure within the mediastinum*: the lungs, as the balloon, tend to collapse *the more the greater their volume*, down to the *residual volume*, the minimum volume the lungs attain within the body, which is greater than the volume where lung's tendency to collapse is nil (see Fig. 3.21). As in the balloon in Fig. 3.30b, the tendency of the lungs to collapse *causes a negative pressure in the tissues around them, i.e. in the mediastinum* (Donder's negative pressure).

The lungs are attached to the thoracic wall by the adhesion of their visceral pleura to the parietal pleura attached to the surrounding structures (thorax wall and mediastinum). The intrapleural pressure is the negative pressure of the liquid filling the small virtual spaces between visceral and pleural surfaces. Liquid in the intrapleural space is continuously absorbed by the lymphatic vessels in the thorax wall and by the oncotic pressure in the alveolar capillaries (see Sect. 1.7.2). If air is allowed to enter between the two pleural sheets the lung collapses and a *pneumothorax* ensues. If liquid is injected a *hydrothorax* is obtained. In both cases the pressure between the two pleural sheets equals $-P_L$, i.e. the tendency of lung to collapse. If air or liquid are extracted, the visceral pleura will approach the parietal one until one contact the other, i.e. the *shape* of the lung surface must fit that of the thoracic wall. This will ultimately lead to small curvatures of the visceral pleura whose elastic retraction causes a negative intrapleural pressure opposing further adsorption of liquid by the lymphatic vessels and the oncotic pressure of the alveolar capillaries. The intrapleural pressure is more negative then that due to the tendency of the lungs to collapse — P_1, with the consequence that the lungs remain expanded, 'attached' to the parietal pleura.

3.7.3 Pressure-Volume Diagram of the Combined System Lungs-Chest Wall: The Relaxation Curve

Figure 3.31 shows the experimental procedure followed to determine the *relaxation curve* P_{rs}, i.e. *the alveolar air pressure when the subject relaxes his respiratory muscles at a given lung volume, with its two components: the pressure volume diagrams of the lungs P_l and of the chest wall P_w* (Fig. 3.32). The subject breathes into a spirometer (as that described in Fig. 3.31), with a balloon in his esophagus (not drawn in the Figure) connected to a water manometer to measure the mediastinal pressure, i.e. the pressure of the alveolar air minus the pressure due to the tendency of the lungs to collapse P_1. Between the subject and the spirometer a three ways faucet allows disconnecting the subject from the spirometer and to connect his mouth to a water manometer. The subject breathing normally is invited to take a maximal inspiration attaining 100% of his vital capacity; at this point the faucet is turned disconnecting the subject from the spirometer (horizontal lines in the spirometric chart on the right of Fig. 3.31) and the subject is invited to relax his respiratory muscles against the manometer measuring the alveolar relaxation pressure (P_{rs} in Fig. 3.32). After this maneuver the connection towards the spirometer is reestablished and the subject breathes normally until the test is repeated at different values of %CV to obtain the full *relaxation curve* (continuous line in Fig. 3.32). The two components, P_w and P_l, of the relaxation curve, P_{rs}, are obtained as follows: since $P_{med} = P_{rs} - P_l$ and $P_{rs} = P_l + P_w$, it follows that $P_{med} = P_l + P_w - P_l = P_w$ and $P_l = P_{rs} - P_w$.

Figure 3.32 shows that the volume where equilibrium is attained between tendency of the lungs to collapse P_1 and the chest wall to expand P_w corresponds to 35% of the vital capacity in a sitting upright normal subject. This is the volume at the end of a normal expiration at rest. If the tendency of the lungs to collapse decreases due to a reduction of their elastic tissue (emphysema), as physiologically tends to occur in the elderly, the volume of the lungs at rest increases (barrel-shaped chest). Note that the equilibrium volume of the chest wall is 55% of the vital capacity; above this value the chest wall, as the lungs, tend to collapse attaining together a pressure of 50 cm of water at 100% of the vital capacity (10 cm H_2O the chest wall, 40 cm H_2O the lungs). After a maximal expiration down to 0% of vital capacity the thorax-lung system tends to expand with a pressure of -45 cm of water mostly due to the thorax wall tendency to expand (-47 cm H_2O) with only a small pressure to collapse of the lungs at the residual air volume ($+2$ cm H_2O).

Figure 3.33 shows the effect of gravity on the mechanics of the respiratory system. When the subject is standing upright, gravity pulls down the diaphragm increasing the equilibrium volume of the chest wall (exceeding the smaller expiratory effect due to downward rotation of the ribs). In a subject laying in bed this condition is reversed: the weight of the abdominal bowels pushes the diaphragm up decreasing the equilibrium volume relative to the standing position. Note that gravity changes the pressure-volume relation of the chest wall only, leaving unaltered the pressure-volume relation of the lungs contained in it. It follows that the equilibrium position of the whole system, where the tendency of the chest wall to expand equals the

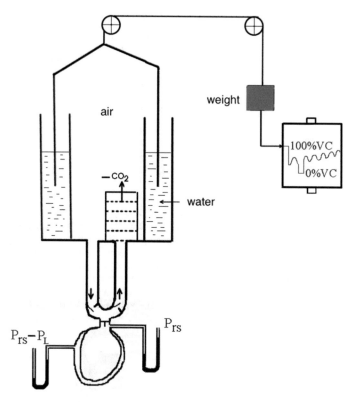

Fig. 3.31 Apparatus used to determine the graphs in Fig. 3.32. Same indications as in Fig. 3.17 except for: (i) the two balloons indicating the chest wall, the external one, and the lungs, the internal one; (ii) the water manometer on the right measuring the alveolar pressure P_{rs} when the three ways faucet is closed, impeding lung-spirometer connection, and the respiratory muscles are relaxed against the water manometer (horizontal tracts in the chart record to the right); (iii) the water manometer on the left measuring the mediastinal pressure, $P_{med} = P_{rs} - P_l$, (where P_l is the tendency of the lung to collapse)

tendency of the lungs to collapse, is lower in the supine position (about 22% of CV in Fig. 3.33). The alveoli and the airways of a patient lying in bed will be therefore less expanded than in the upright position i.e. more prone to closure (partial lung atelectasis). The diameter of the intrapulmonary airways depends on two factors: (i) a *static* factor increasing their diameter with lung volume due to the pull on their walls by the surrounding alveoli; (ii) a *dynamic* factor due to the different air pressure across their walls, which tends to compress them during expiration and to dilate them during inspiration. The *static* factor decreases airways diameter in emphysema due to partial destruction of the surrounding tissue, with the consequence that the patient to prevent airways collapse decreases the airflow during expiration by partial closure of his lips.

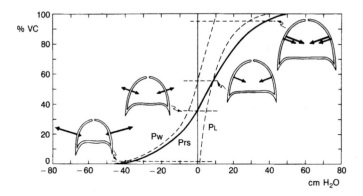

Fig. 3.32 The continuous line shows the relaxation curve P_{rs} of human combined lungs-thorax ensemble expressed as a % of the vital capacity VC as a function of the alveolar pressure in sitting position. The interrupted lines indicate its two components: P_L, the tendency of the lungs to collapse at all %CV values, and P_w, the tendency of the thorax to expand up to about 55% of CV and to collapse at greater volumes. The arrows in the schematic drawings indicate the pressures exerted by the lungs and by the thoracic wall at the volumes indicated by the horizontal interrupted lines (from Knowles JH, Hong SK and Rahn H *J Appl Physiol* 14: 525, 1959, as modified from Agostoni E and Mead J *Statics of the respiratory system* in Fenn WO and Rahn H (eds) *Handbook of Physiology*, Am. Physiol. Soc., Washington DC, 1964)

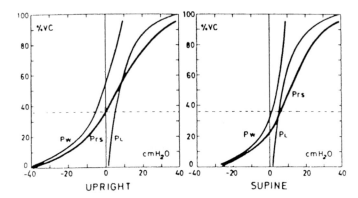

Fig. 3.33 Effect of posture on the relaxation curve P_{rs} showing that the equilibrium volume attained at 35% of vital capacity CV in the upright position (horizontal interrupted line) decreases to about 22% when lying down. Other indications as in Fig. 3.32 (from Agostoni E and Mead J *Statics of the respiratory system* in Fenn WO and Rahn H (eds) *Handbook of Physiology*, Am. Physiol. Soc., Washington DC, 1964)

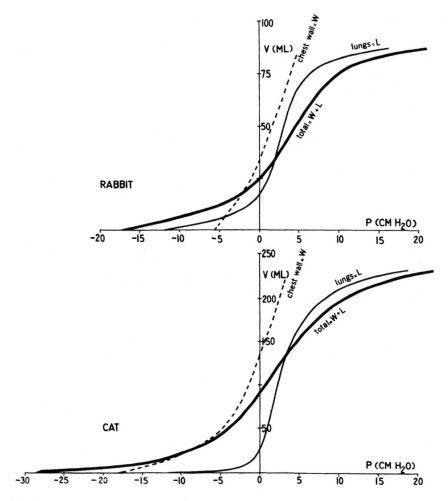

Fig. 3.34 Relaxation curves measured down to zero lung volume (complete atelectasis) (from Cavagna GA, Stemmler EJ and DuBois AB *J. Appl. Physiol.* 22: 441–452, 1967)

In Fig. 3.34 the relative contribution of the lungs and of the chest wall to the total relaxation pressure are shown down to zero lung volume (not 0% CV as in Figs. 3.32 and 3.33). It appears that at zero volume (intercept on the abscissa) a considerable fraction of the negative gradient of pressure across the thorax is sustained by the lungs; and in fact in the rabbit the lungs resist collapse more than chest does.

3.7.4 Maximal Expiratory and Inspiratory Pressures

The experimental set up used is similar to that illustrated in Fig. 3.31 except that the manometer at the mouth is filled with mercury and no mediastinal pressure is measured. The subject is invited to make maximal expiratory and inspiratory efforts against the mercury manometer at different lung volumes measured as % of the vital capacity CV (Fig. 3.35). At 100% of CV the maximal pressure exerted with a maximal expiratory effort is ~100 mmHg; since $P\,V = n\,R\,T$ = constant, the increase in pressure causes a reduction in volume of the air in the lungs as indicated by the negative slope of the upper arrow in Fig. 3.35; the maximal inspiratory pressure at 100% of CV is obviously nil because the maximal effort has already be done by the inspiratory muscles just to *attain* 100% CV. The contrary is true at 0% CV where the maximal negative pressure is about −90 mmHg. The interrupted line above 100% CV refer to a condition where air is pumped within the lungs causing their rupture. Vice versa sucking air from the lungs below 100% CV will cause a dilatation of the pulmonary vessels. The pressure-volume area between the maximal expiratory and inspiratory curves represent the maximal work that can be done by the respiratory muscles ($P \times V = F/L^2 \times L^3 = F \times L$).

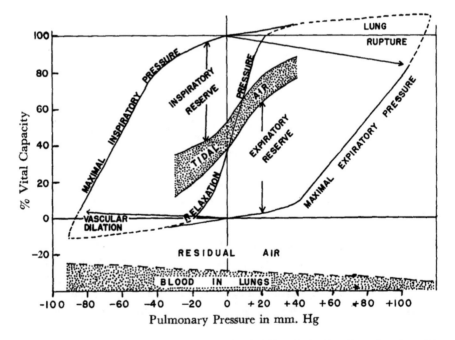

Fig. 3.35 The lung volume expressed as % of the Vital Capacity is plotted as a function of the alveolar pressure in different experimental conditions as indicated and explained in the text (from Fenn WO *Am J Med* 10: 77–90, 1951)

The *tidal-air* dotted area in Fig. 3.35 shows the respiratory tidal volume V_t at different trans pulmonary pressures, i.e. the amplitude of each breath, of an awake subject breathing against a positive air pressure (right: e.g. a kid treated for hyaline membrane disease) and negative air pressure (left: e.g. a subject under water breathing air on the surface through a pipe). During normal breathing when the average pressure within the lungs equals that outside (zero on the abscissa of Fig. 3.35) the *inspiration is active*, i.e. lung volume is increased by muscular contraction above the point where the relaxation curve crosses the ordinate (35% of CV), and the *expiration is passive*. The same is true when the gradient of pressure across the lungs is negative, whereas the opposite is true when the pressure on the abscissa of Fig. 3.35 exceeds + 10 mmHg. A curarized subject will attain at each pressure on the abscissa the volume indicated by the relaxation curve on the ordinate: ±20 mmHg are enough to attain the 100% and the 0% of CV! If a subject is 130 cm below the surface of the water with his lungs connected with a pipe to the surface of the water his chest wall will be compressed with $1300/13.6 = 96$ mmHg, i.e. he will be barely able to maintain ~0%CV with a maximal inspiratory effort!

3.8 Respiratory Work

3.8.1 *Graphic Representation*

Figure 3.36 shows the *positive inspiratory work* that the inspiratory muscles must due at each breath during normal respiration in an upright position. The inspiration begins from the point of equilibrium where the tendency of the lungs to collapse equals the tendency of the chest wall to expand (zero alveolar pressure at 35%CV). In order to introduce air into the lungs the inspiratory muscles must exert a negative pressure in the alveoli, which is greater the greater the airflow. In the schema of Fig. 3.36 the airflow (nil at the beginning and at the end of inspiration) is assumed to attain a maximum at half V_t. The deep gray area between the curve so obtained and the ordinate represents the *dynamic work* done during the inspiration and will be greater the greater the velocity of the airflow. Now assume to introduce air so slowly that the negative pressure on the abscissa remains practically equal to zero. Even in this case the inspiratory muscles must do work to expand the chest wall because the increase in volume will increase the tendency to collapse of the chest wall-lungs system as indicated by the relaxation curve *Prs*. This work, that we can call *static work*, is indicated by the light gray area in Fig. 3.36. For a given V_t the static work is always the same whereas the dynamic work increases with the velocity of the airflow within the lungs.

Figure 3.37 shows the expiratory work: the light gray area is *passive* positive work sustained by the elastic recoil of the chest wall-lung system *Prs*; the dark gray area indicates positive work done by *active* contraction of the expiratory muscles exerting a pressure exceeding that due to elastic recoil; the interrupted lines trian-

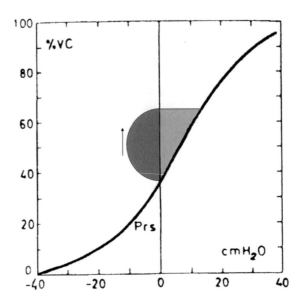

Fig. 3.36 Inspiration: the lung volume expressed as a % of the vital capacity, CV, is plotted as a function of the alveolar air pressure in centimeters of water. The thick line indicates the relaxation curve in upright position (left graph in Fig. 3.33). The dark gray area indicates the dynamic *inspiratory* work to sustain the flow of air into the lungs. The light gray area indicates the static elastic inspiratory work required to expand the wall thorax-lungs system

gle represents *negative* work done by the inspiratory muscles which do not relax immediately and simultaneously after the end of inspiration. *Smaller airflow rate during expiration would be represented by a light gray area totally contained within the triangle between ordinate and the Prs curve: in this case expiration would be completely passive, i.e. sustained by the elastic energy stored during inspiration.* When the ventilation exceeds ~20 l/min *the dynamic work represents the total respiratory work* because the positive elastic work done during inspiration is stored as elastic potential energy and recovered during expiration. Below this ventilation (as in Fig. 3.39) the elastic potential energy acquired during the expansion of the lungs is sufficient to sustain the dynamic work done during expiration, i.e. the total work equals the work done during inspiration: the expiration is passive.

On the opposite case, maximal and fast V_t would require a work approaching the area between the maximal inspiratory and expiratory pressure curves of Fig. 3.35.

3.8.2 Arithmetic Representation

Assuming that the *Prs* curve is a straight line, the *static* average pressure exerted during inspiration to overcome the elastic recoil will be $\overline{P} = kV_t/2$ (considering that

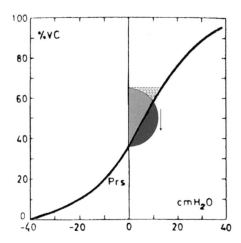

Fig. 3.37 Expiration: ordinate, abscissa and relaxation curve as in Fig. 3.36. The light gray area represents the passive expiratory work sustained by the elastic energy stored within the thorax wall-lungs structures during the inspiration. The dark gray area represents the active positive work done by the expiratory muscles to sustain airflow during expiration. The interrupted lines area represents the *negative* work done by the *inspiratory* muscles opposing immediate recoil of the thorax wall-lungs structures at the beginning of expiration

inspiration starts from the point where the *Prs* curve crosses the ordinate, Fig. 3.36). As shown in Sect. 3.4, the *dynamic* pressure P propelling a fluid is $P = a\dot{V} + b\dot{V}^2$ where the first term refers to the laminar motion and the second term to the turbulent motion of the flow $\dot{V}$. Assuming a sinusoidal path with period τ of the spirometric record and $V_{ti} = V_{te} = V_t$ (as in Figs. 3.8 and 3.13), the *average flow during inspiration* will be $V_t/(\tau/2)$, laminar, and $[V_t/(\tau/2)]^2$, turbulent, and the average pressure: $\overline{P} = kV_t/2 + aV_t/(\tau/2) + b[V_t/(\tau/2)]^2 = kV_t/2 + 2af\,V_t + 4bf^2V_t^2$ (note that $f = 1/\tau$), the first term of $\overline{P}$ is the *static* component whereas the second and the third terms represent the *dynamic* components: laminar and turbulent. The *work* (pressure time volume) will be $W = \overline{P} \cdot V_t = kV_t^2/2 + 2af\,V_t^2 + 4bf^2V_t^3$ and the *power*:

$$\dot{W} = W \cdot f = kf\,V_t^2/2 + 2af^2V_t^2 + 4bf^3V_t^3$$

The three terms of this equation: elastic, laminar (viscous), turbulent and their sum (tot.) are plotted in Fig. 3.38 as a function of the breathing frequency for the *same alveolar ventilation*. At the end of Sect. 3.5.2, the conclusion was reached that disregarding the mechanical work done to sustain respiration which will be treated later, deep breaths at a lower frequency will be more efficient than shorter breaths at a high frequency maintaining the same pulmonary ventilation. Figure 3.38 shows that taking into account the mechanical work done to sustain respiration an *'optimal' frequency exists where mechanical power attains a minimum for the same alveolar ventilation*. If an elastic bandage is wrapped around the chest, the elastic and the total power are increased with the result that the 'optimal' breath frequency is

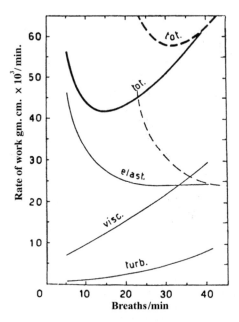

Fig. 3.38 The mechanical inspiratory power to maintain the same alveolar ventilation is plotted on the ordinate as a function of the frequency of breathing. The thinner lines indicate from bottom to top the power spent to sustain the turbulent, laminar (visc.) motion of air and to expand the wall-lungs thorax structures (elast.). The thicker lines indicate the total power output. The interrupted lines indicate a condition where the elastic power is increased by wrapping an elastic bandage around the chest. Note that power attains a minimum at a given frequency, which is higher the stiffer the system (modified from Otis AB, Fenn WO and Hermann R *J Appl. Physiol.* 2: 592–607, 1950)

increased (interrupted lines in Fig. 3.38). Note that the elastic *static* power increases sharply when alveolar ventilation is maintained with deep breath at a lower frequency whereas the contrary is true for the *dynamic* power.

3.8.3 *Esophageal Pressure*

In Fig. 3.39a the esophageal pressure is plotted as a function of time during a single breath (from the end of expiration *e* to the end of inspiration *i*). The interrupted line is a segment of the P_L curve in Fig. 3.33 showing the *static* negative mediastinal pressure due to the elastic tendency of the lungs to collapse. The thicker line shows the sum of the static pressure P_L and the *dynamic* pressure required to sustain the flow of air and viscosity structures of the lung tissues during inspiration (*e* to *i*) and expiration (*i* to *e*).

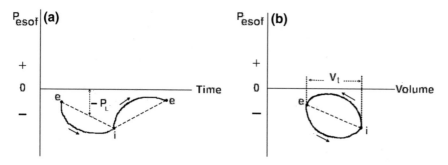

Fig. 3.39 a The esophageal pressure (i.e. the mediastinal pressure around the heart) is plotted as a function of time during inspiration (*e* to *i*) and expiration (*i* to *e*); the interrupted line shows the negative pressure due to the *static* elastic recoil of the lungs ($-P_L$, see Fig. 3.33), the continuous line indicates the sum of the static pressure and the *dynamic* pressure due to airflow and lung viscosity. Note that the dynamic pressure is negative during inspiration, i.e. adds to the static pressure, whereas it is positive during expiration, i.e. subtracts from the static pressure: higher flow rates result in positive mediastinal (esophageal) pressure during expiration. **b** The esophageal pressure is plotted as a function of the volume change during one breath of amplitude V_t. In this case the interrupted line represents a part of the *P-V* diagram of the lungs (Fig. 3.33) and the area between the interrupted line and the abscissa represent the static elastic work done to expand the lungs during inspiration (*e* to *i*) and, in absence of hysteresis loss, recovered during expiration (*i* to *e*). *The area between the two thicker lines represents the dynamic work done to sustain the motion of the air plus the lungs* (excluding chest wall)

Figure 3.40 shows the mean mediastinal pressure during a respiratory cycle $P_m = A/(2V_t)$, where A is the area of the respiratory loop (Fig. 3.39b) and V_t the tidal volume, as a function of the mean airflow $\dot{V}_m = (2V_t/\tau)$ during the breath (*inspiration plus expiration*, broken line). The continuous line was calculated from values of pulmonary ventilation (V_t/τ) greater than 45 l/min ($\dot{V}_m > 1.5$ l/s) assuming that hysteresis is inappreciable, and that the function is well defined by the equation: $P = a\dot{V} + b\dot{V}^2$. For values of $\dot{V}_m$ greater than 2 l/s the two curves are almost superimposed, whereas at low flow values, the experimental values are higher than those calculated as $P = a\dot{V} + b\dot{V}^2$: the function is better defined by $P_m = i + a'\dot{V}_m + b'V_m^2$ calculated using all the experimental data, where i expresses the mean pressure needed to overcome pulmonary hysteresis (0.4–0.5 cmH$_2$O). For a normal rest flow, $\dot{V}_m = 0.3$ l/s, the mean pressure to overcome lung hysteresis is about equal to the pressure necessary to overcome airway resistance and lung viscosity.

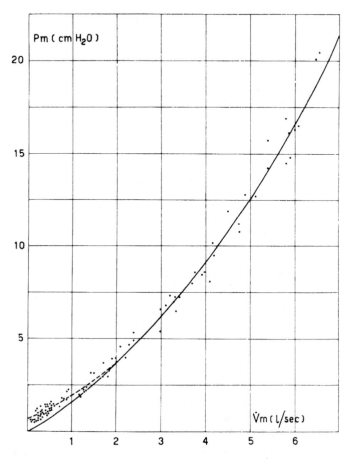

Fig. 3.40 Mean pressure (P_m) plotted as a function of the mean flow $\left(\dot{V}_m\right)$. Continuous line is drawn according to function $P = a\dot{V} + b\dot{V}^2$, a and b being calculated from data obtained at $\dot{V}_m > 1.5$ l/s. Broken line is calculated according to function $P_m = i + a'Vm + b'Vm^2$ using all the experimental data. For values of $\dot{V}_m > 2$ l/s the two curves are almost superimposed (from Cavagna GA, Brandi G, Saibene F and Torelli G, *J. Appl. Physiol.*, 17: 51-53 1962)

3.9 Diffusion

Communication between internal and external surrounding of our body (essential to maintain constant the physics and the chemistry of our organs) takes place in two ways: one appropriate for large distances (e.g. blood circulation and respiration), and one working for small distances: the *diffusion* (as taking a bus and then walking from bus stop to home). Diffusion is an example of the tendency of every system to attain its most probable condition, i.e. a minimum of potential energy. For this reason any group of molecules subject to a Brownian motion will spontaneously expand from a site of greater concentration to a site of smaller concentration up to a condition where

Fig. 3.41 Unidirectional diffusion of oxygen within a tube of area A along the direction (modified from Ackerman E *Biophysical Science*, Prentice Hall Inc., 1962)

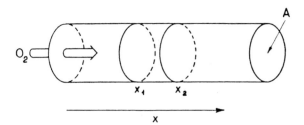

concentration is equal in all sites. Molecules may diffuse in a gaseous environment, in a liquid environment and even in a solid environment. For example a volume of oxygen at the extremity of a tube containing nitrogen will spontaneously diffuse towards the other extremity of the tube until the condition is reached where oxygen concentration is equal all along the tube. At the level of the capillaries gases diffuse within a liquid. Diffusion may also occur within a solid surrounding (e.g. mercury within gold). The *velocity* of diffusion is maximal in a gaseous environment, smaller in a liquid environment and minimal in a solid environment. Let's derive logically the law governing diffusion (Fick principle) (Fig. 3.41).

The total mass of oxygen molecules subjected to Brownian motion, which may move in all directions in unit time will be greater the greater the area A they occupy at the position x_1 and the greater their concentration C_1 over that area, i.e. $dm_1/dt = \beta A C_1$. Similarly at the position x_2, $dm_2/dt = \beta A C_2$. The net transfer of molecules from x_1 to x_2 will be $dm/dt = \beta A (C_1 - C_2)$. Let's now imagine a graph where C is plotted as a function of x; as shown below the slope of this graph $\Delta C/\Delta x$ is a negative number:

$$C_0 > C_1 > C_2 > C_3 > C_4 \qquad C_2 - C_1 = \Delta C \text{ is negative}$$

$$X_0 < X_1 < X_2 < X_3 < X_4 \qquad X_2 - X_1 = \Delta x \text{ is positive}$$

It follows that the net transfer of molecules from x_1 to x_2 defined above: $dm/dt = \beta A(C_1 - C_2)$ must be written as $dm/dt = -\beta A \Delta C$. The mass of oxygen diffusing from one area to the adjacent area is logically greater the smaller the distance between the two areas Δx. For this reason the proportionality constant β has been assumed to be inversely proportional to Δx, i.e. $\beta = D/\Delta x$ so that the *general equation governing gas diffusion (the first Fick's law) is* $dm/dt = -D A \Delta C/\Delta x$, and D is called *coefficient of diffusion*, having dimensions $[l^2 t^{-1}]$, units cm^2/s, and *for several gasses is inversely proportional to the square root of the molecular weight* and depends on the medium where diffusion takes place. The coefficient of diffusion of oxygen in air is ~0.18 cm^2/s similar to the diffusion of *heat* in air (0.20 cm^2/s).

Fig. 3.42 Diffusion of
oxygen against a
concentration gradient
(modified from Forster RE in
Fenn WO and Rahn H (eds.)
Handbook of Physiology,
Sect. 3, Vol. I, American
Physiological Society,
Washington DC, 1964)

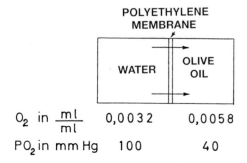

	WATER	OLIVE OIL
O_2 in $\frac{ml}{ml}$	0,0032	0,0058
PO_2 in mm Hg	100	40

Figure 3.42 shows one case where oxygen diffuses from a medium (water) where its concentration is lower towards a medium (oil) where its concentration is higher in apparent contrast with Fick's law. This is because oxygen is *physically dissolved* in water and oil. What means 'physically dissolved'? This question introduces a new concept, i.e. the *Henry*'s law stating that the concentration of a gas dissolved in a liquid is proportional to the partial pressure of the gas, i.e. *ml gas/ml liquid* $= \alpha\, pp_{gas}$; the proportionality constant α is the *coefficient of solubility* of the gas which is greater the cooler the liquid. The apparent contrast of the example in Fig. 3.42 with Fick's law shows that *Fick's law applies when diffusion takes place in a single phase*. In our organism therefore, where several contiguous phases are present, diffusion takes place following a gradient of *partial pressure*, which we can imagine as the tendency of molecules to 'escape'. The concentration $C = \alpha\, pp$, i.e. $\Delta C = \alpha\, \Delta pp$. *Therefore we can rewrite Fick's law as: dm/dt* $= -D\,A\,\alpha\,\Delta pp/\Delta x$. It follows that molecules with similar D (e.g. O_2 and CO_2) may diffuse differently if the coefficient of solubility α of one molecule is different from that of the other molecule; for example CO_2 is more soluble than O_2 in water and therefore diffuses more quickly than O_2 in spite of a similar D.

In practice (e.g. Hospital, Clinic) the *Fick's law: dm/dt* $= -D\,A\,\alpha\,\Delta pp/\Delta x$ is written: $\Delta m/\Delta t\,\Delta p = D\,\alpha\,A/\Delta x$ and, considering the oxygen consumption, the *capacity of diffusion of the lung D_1 is defined as*: $D_1 = \dot{V}O_2/\Delta p = D\alpha\,A/\Delta x$ (actually D_1 is a *conductance*, i.e. the reciprocal of the *resistance* $1/D_1$ that the molecules of oxygen must overcome to reach the hemoglobin). In clinical terms A is the total surface of the lungs at disposal for gas exchange (which may be reduced in case of atelectasis or pulmonary resection), and Δx is the thickness, the barrier, through which oxygen must diffuse (which may be increased in case of pulmonary edema).

Let's now consider the journey that the molecule of oxygen must travel to reach the molecule of hemoglobin from the ambient air. During inspiration the first 150 ml of air in the dead space entering into the alveolar air volume have the same composition of the alveolar air; it follows that no diffusion of oxygen occurs towards the preexisting functional residual air volume. The other 350 ml of fresh ambient air (assuming $V_t = 500$ ml at rest), having a greater concentration of oxygen, mix with the preexisting functional residual air volume. The question arises: oxygen travels to the alveolar membrane by diffusion or by mechanical mixing?

In order to answer this question two experiments were made. In the first experiment, a subject was invited to inspire V_{ti} of air mixed with helium in one case and with an aerosol in the other case. After a pause the subject expires and an analysis is made of the contents of his V_{te} volume: it was found that most of the aerosol particles were found in V_{te} whereas the concentration of helium was drastically reduced. This experiment shows that the gas into gas travel from the ambient air to the alveolar membrane is sustained by diffusion and not by mechanical mixing (which is higher for the aerosol than for helium given the greater kinetic energy of the larger mass of aerosol particles vs. helium molecules).

The other experiment is made using a plethysmograph. It was found that the pulmonary volume, measured as described in Sect. 3.5.2 assuming a constant temperature during compression of the air in the lungs, is the same up to a frequency of 150/min in-expiratory efforts against a manometer. This means that temperature equilibrium is attained in $60\ s/300 = 0.2$ s. Note that, as mentioned above, the coefficient of diffusion of oxygen in air is similar to the diffusion of heat in air ($\sim 0.20\ cm^2/s$). In conclusion, the gas into gas travel by diffusion is so fast ($\sim 200{,}000$ times that in tissues and blood) that its resistance is negligible compared with that of gas into liquid (from the alveolar membrane to the molecule of hemoglobin, see below) even if the distance of the gas into gas travel ($\sim 150\ \mu m$) is greater than that from the liquid surface of alveoli to hemoglobin ($\sim 3.5\ \mu m$).

These experiments show that the resistance to diffusion of inspired air into the pre-existing functional residual capacity volume can be neglected. After this travel, the molecule of oxygen must overcome the *resistance* offered by *the pulmonary membrane, which is physiologically represented by all the phases interposed between alveolar air and molecule of hemoglobin*: (i) the *oriented* molecules of the tension active material (forming a film which offers more resistance to diffusion than the tension active molecules in solution), (ii) the epithelium, (iii) the endothelium, (iv) the plasma and (v) the distance to cover within the red cell before reaching the hemoglobin molecule (Fig. 3.43). The ensemble of all these resistances is physiologically represented as $1/D_m$ where $D_m = $ ml gas/(min $\times$ mmHg). In this surrounding the velocity of diffusion of the oxygen molecules is much smaller than in the gas environment because the molecules of the medium where O_2 has to diffuse are packed against each other. After overcoming the resistance $1/D_m$ the molecule of oxygen faces the heme of the hemoglobin molecule. Even if with a different mechanism a gradient of oxygen molecules builds also in front of the heme due to the limited velocity of reaction of the oxygen binding to heme (as a queue of persons in front of a front office). This 'resistance' is defined as $1/\theta V_c$ where $\theta = $ ml gas/(min $\times$ mmHg $\times$ ml *of blood*) and V_c are the ml of blood in the lungs to which oxygen binds. The problem is that after combining with hemoglobin the oxygen tends to dissociate from it with a *back pressure* described by the blood-oxygen dissociation curve (which we will describe later on), i.e. $HbO_2 \rightarrow Hb + O_2$. As described schematically below, the measure of *the total resistance* $1/D_1 = 1/D_m + 1/\theta V_c$ is complicated by the lack of knowledge of the back pressure.

The *initial* pO_2 (on the left side of Fig. 3.44) is easily determined by analyzing the air at the end of expiration (i.e. after mixing with the inspired air), but how about

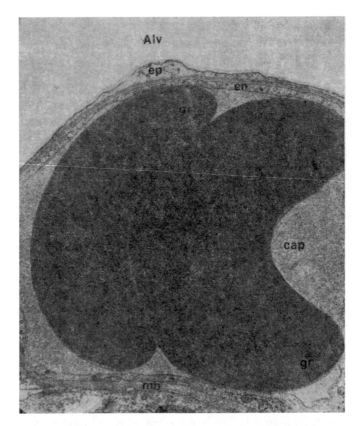

Fig. 3.43 Section of a blood capillary (*cap*) contained on the alveolar wall of a human showing the path that gas must travel by diffusion between alveolar air (*Alv*) and red cell (*gr*), a path consisting of *ep*, the alveolar epithelium; *en* the endothelium and the plasma around the red cell; *mb* is the basal membrane (courtesy of De Gasperis C, Institute of Human Anatomy, University of Milan)

the measure of the *final* pO_2 within the heme, which increases with Hb oxygenation according to a complicated function (Fig. 3.45)? On the other hand, as described above, the measure of $D_l = \dot{V}O_2/\Delta p = D\alpha A/\Delta x$ is useful in clinic because it gives indication of the area A and thickness Δx of the alveolar surface of a patient. To solve this problem one uses instead of O_2, small volumes of carbon monoxide CO, whose affinity for hemoglobin is 230 times greater than that of oxygen, with the consequence that its backpressure can be neglected. It follows that $D_{l,CO} = \dot{V}CO/Alveolar\ pressure$ since the pressure of CO within the heme is ~zero. The relation we have to consider now on is therefore the sum of two resistances to the diffusion of CO: that through the membrane and that in front of the heme: $1/D_{l,CO} = 1/D_{m,CO} + 1/\theta_{CO}V_c$.

At saturation with O_2, blood binds 20 ml of O_2/100 ml of blood, in an *adult* the total blood volume is 5 l binding a total of 1 l of O_2. If we give to the subject 1 l of

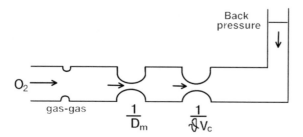

Fig. 3.44 Schematic representation of the resistances that a molecule of oxygen must overcome: (i) to travel within the air of the functional residual capacity (*gas-gas*: negligible); (ii) through all phases interposed between air and the hemoglobin molecule ($1/D_m$: see Fig. 3.43) and (iii) to bind to the heme of the hemoglobin molecule ($1/\theta V_c$). The vertical tube on the right represents the tendency of the oxygen molecules to flow backwards after binding to the hemoglobin molecule

Fig. 3.45 Showing how oxygen *back pressure* P_1 increases progressively during its course through the pulmonary capillary making it difficult to assess the average $\Delta p = P_2 - P_1$ in the measure of D_{lO_2} (after Kety SS, *Methods in medical research*, Vol. II)

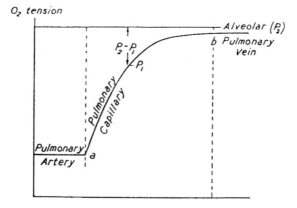

CO, we kill him because no more O_2 could be available for his metabolism, 100 ml of CO will be a lot, 10–50 ml of CO or less could be considered according to the clinic condition of the subject.

The equation $1/D_{l,CO} = 1/D_{m,CO} + 1/\theta_{CO}V_c$ can be represented graphically as a straight line with $1/D_{l,CO}$ on the ordinate (measured in vivo as $D_{L,CO} = \dot{V}CO/Alveolar\ pressure$) and $1/\theta_{CO}$ on the abscissa, determined in vitro after mixing a CO solution with a solution of *hemoglobin combined with different values of partial pressure of oxygen*. The greater the partial pressure of oxygen P_{O_2} (top of Fig. 3.46), and as a consequence the concentration of HbO_2, the smaller the affinity θ_{CO} with Hb and the greater $1/\theta_{CO}$ (abscissa of Fig. 3.46). This very interesting experiment allows to determine the resistance offered by the alveolar 'membrane' $1/D_{m,CO}$ as the intercept of the straight line with the ordinate and the volume of the blood in both lungs as the slope of the line. The intercept shows that at the normal partial pressure of oxygen in the alveoli $P_{O_2} \sim 100$ mmHg the resistance offered by the 'membrane' $1/D_m$ is about two times greater than that to bind to hemoglobin; the slope shows a volume of pulmonary blood of only 100 ml; this small volume is distributed over

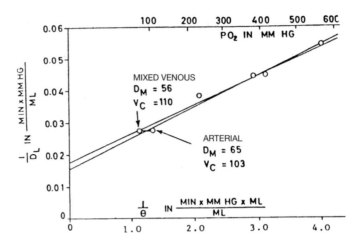

Fig. 3.46 The total resistance to the diffusion of CO into the lungs ($1/D_{l,CO}$, ordinate, determined in vivo) is plotted as a function of the resistance to the diffusion of CO into 1 ml of blood ($1/\theta_{CO}$, abscissa, determined in vitro). Measurements are made at different partial pressures of oxygen PO_2 (abscissa at the top of the Figure) resulting in different values of $1/\theta_{CO}$. The value of $1/D_{l,CO}$, corresponding to the lowest value of PO_2, is given for values of $1/\theta_{CO}$ corresponding to the value of PO_2 at the beginning (*mixed venous*) and at the end (*arterial*) of the alveolar capillary: the correct value must be between the two. The two straight lines are calculated by the least squares method taking into account each of the two points; the corresponding values of D_m (reciprocal value of the intercept) and of V_c (reciprocal value of the slope) are indicated in the Figure (from Roughton FJW and Forster RE, *J Appl. Physiol.* 11:291, 1957)

an alveolar surface of about 70 m^2 indicating how thin is the thickness of the blood layer.

3.10 Breathing at High Altitude

As described in the Introduction (Fig. 3.3), barometric pressure P_b decreases exponentially with increasing altitude and since the fraction of oxygen in air remains constant ($F_{i,O_2} = 20.93\%$) its partial pressure $P_{i,O_2} = F_{i,O_2} \times P_b$ decreases with altitude. The same mass of air molecules contained in a $V_t = 500$ ml will occupy 1 l at $P_b/2$ (6500 m), i.e. the amplitude of our breaths will increase resulting in a greater water loss by evaporation which may lead to dehydration.

3.10.1 Maximal Altitude Attainable by Humans

The partial pressures of gas and water vapor at sea level are given in mmHg in Table 3.5. It can be seen that the total pressure amounts to 760 mmHg except for

Table 3.5 From Pittman RN *Regulation of tissue oxygenation*, San Rafael (CA), Morgan and Claypool Life Sciences, 2011

	Dry air	Moist tracheal air (37 °C)	Alveolar gas	Arterial blood	Mixed venous blood
P_{O_2}	159.1	149.2	104	100	40
P_{CO_2}	0.3	0.3	40	40	46
P_{H_2O}	0.0	47.0	47	47	47
P_{N_2}	600.6	563.5	569	573	573
P_{total}	760.0	760.0	760	760	760

the venous blood where it attains 706 mmHg only due to the fact that the oxygen pressure falls from arterial blood to venous blood more than the carbon dioxide pressure increases (the reason will be explained later).

The maximal altitude attainable when *breathing pure oxygen* has been determined in laboratory condition on a normal non-acclimatized subject by measuring the minimal partial pressure of oxygen compatible with his consciousness (his writing capability). It was found that the minimal alveolar pressure of oxygen was P_{alv,O_2} = 30 mmHg and the minimal P_{alv,CO_2} = 20 mmHg (reduced from 40 mmHg because of hypoxia induced hyperventilation). Taking into account that the partial pressure of water vapor at saturation and at 37 °C, as in the alveoli, is always P_{alv,H_2O} = 47 mmHg, independent of the altitude, the total pressure compatible with consciousness was found to be 30 + 20 + 47 = 97 mmHg corresponding to an altitude of ~14,000 m. When *breathing air* we must add the partial pressure of nitrogen P_{N_2}; if the respiratory quotient $QR = CO_2/O_2 = 1$, the fraction of P_b occupied by the *respiratory gases* ($CO_2 + O_2$) remains 21% even at low P_b values; it is therefore possible to calculate the partial pressure of nitrogen in the hypoxia condition described above from the following proportion: 50 mmHg $\left(P_{CO_2} + P_{O_2}\right)$: 21% = P_{alv,N_2}: 79%, i.e. P_{alv,N_2} = 188 mmHg. The minimal barometric pressure P_b breathing air will then be 30 + 20 + 47 + 188 = 285 mmHg, corresponding to 7500 m above sea level. However in 1978 Reinhold Messner climbed to the top of Mount Everest, 8848 m, breathing air without supplemental oxygen! Subsequent analysis on very trained subjects breathing air on top of the Everest found that P_{alv,O_2} = 35 mmHg, P_{alv,CO_2} = 7.5 mmHg corresponding to a pH = 7.7 instead of the normal 7.4; this difference, 0.3 = log 2, shows that the concentration of hydrogen's ions was half the normal due to the great alveolar ventilation. The partial pressure of nitrogen will then be: 42.5 mmHg $\left(P_{CO_2} + P_{O_2}\right)$: 21% = P_{alv,N_2}: 79%, i.e. P_{alv,N_2} = 160 mm Hg and P_b = 35 + 7.5 + 47 + 160 = 249.5 mmHg corresponding to the altitude of Mount Everest (actually the average of all pressures at this altitude is slightly lower: possibly the weight of more dense cold air above Mount Everest zone increases slightly pressure at its top allowing humans to climb it). Obviously this goal was possible for subjects trained at high altitudes with a higher concentration of hemoglobin in their blood. The transfer of oxygen from blood to tissues depends on the difference in oxygen partial pressure independent of hemoglobin concentration; however oxygen partial pressure along the capillary will fall more slowly the greater the hemoglobin concentration; it follows that more oxygen will be delivered to the tissues the greater the concentration of hemoglobin. During hyperventilation the P_{CO2} decreases and with it the most important stimulus to the brain respiratory center located in the medulla oblongata and pons. On the other hand hypoxia stimulates the cells of the carotid and aortic bodies. Kidneys helps reducing the alkalosis due to the decrease in CO_2 concentration by increasing the elimination of bicarbonates ($pH = 6.1 + \log$ ($[HCO_3^-]/\alpha\ pCO_2$) as it will be described in Chap. 4.

3.10.2   Equation of Alveolar Air

When *breathing pure oxygen* the total barometric pressure in the alveoli is: $P_b = P_{alv,CO_2} + P_{alv,O_2} + P_{alv,H_2O}$ (47 mmHg), which can be written as

$$P_{alv,CO_2} = (P_b - 47) - P_{alv,O_2}$$

the equation of the alveolar air when breathing pure oxygen: a straight line ($y = cost - x$) with slope -1 and equal intercepts ($P_b - 47$) on the ordinate and the abscissa; the distance between the intercepts of the parallel lines in Fig. 3.47 decreases with altitude because P_b decreases exponentially with altitude (Fig. 3.3). The parallel lines indicate that by increasing the alveolar ventilation at a given altitude (right hand ordinate) the increase in partial pressure of oxygen equals the decrease of carbon dioxide (as we will see this is not true when breathing air). The ordinate on the right was constructed assuming a metabolism at rest corresponding to a constant production of $CO_2 (\dot{V}CO_2 = 320$ ml/min), i.e. $\dot{V}CO_2 = \dot{V}_{alv} \times F_{alv,CO_2} = $ constant, indicating that $\dot{V}_{alv}$ tends to infinity as F_{alv,CO_2} (ordinate on the left) tends to zero.

The HbO_2 lines starting from the abscissa above: (i) show that the percentage of hemoglobin oxygenated decreases as expected with the reduction of the alveolar oxygen pressure P_{alv,O_2} (abscissa below), and (ii) drift to the left indicating that the same oxygenation can be maintained with lower values of P_{alv,O_2} when P_{alv,CO_2} decreases. This is because, as will be described later, the affinity of hemoglobin with oxygen increases when P_{CO_2} decreases. The thicker line *Norm* indicate the respiratory condition of a non-acclimatized subject breathing oxygen with a mask on an airplane: it can be seen that his alveolar ventilation remains constant when P_{alv,O_2} decreases down to ~60 mmHg with an about constant P_{alv,CO_2} and 90% of oxygenated hemoglobin up to an altitude of 12,000 m. By increasing altitude further, the progressive fall of P_{alv,O_2} stimulates the carotid and aortic bodies to increase ventilation; a limit for consciousness is attained at ~14,500 m with a $P_{alv,CO_2} = $ 20 mmHg and $P_{alv,O_2} = 30$ mmHg.

The *equation of alveolar air when breathing air* is a straight line ($y = cost - b$ x: see its derivation in Margaria R and De Caro L, *Principi di Fisiologia Umana*, Vallardi, Milano, 1977):

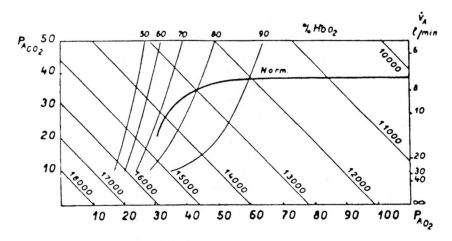

Fig. 3.47 Partial pressure of CO_2 in the alveoli (ordinate) as a function of the partial pressure of O_2 in the alveoli (abscissa) when breathing pure oxygen. The parallel straight lines indicate the equation of the alveolar air for different altitudes in meters. The curve *Norm* indicates the respiratory condition in an untrained subject. The lines starting from the top, iso-HbO_2, indicate the values of P_{alv,CO_2} and P_{alv,O_2} at which the percentage of HbO_2 is constant at the indicated value. The ordinate on the right indicate the alveolar ventilation calculated for a constant production of CO_2 of 320 ml/min (from Margaria R and De Caro L, *Principi di Fisiologia Umana*, Vallardi, Milano, 1977)

$$P_{alv,CO_2} = P_{i,O_2} \times QR/\left[1 - F_{i,O_2}(1 - QR)\right] - P_{alv,O_2} \times QR/\left[1 - F_{i,O_2}(1 - QR)\right]$$

whose slope depends on the ratio between CO_2 produced and O_2 consumed, i.e. on the *respiratory coefficient* $QR = CO_2/O_2$ (Fig. 3.48). It can be seen that when $R = 1$ and when $F_{i,O_2} = 1$, the equations breathing pure oxygen and air coincide and the slope of their straight lines equals $-45°$. When alveolar ventilation is ∞, $P_{alv,CO_2} = 0$ and $P_{i,O_2} = P_{alv,O_2}$ independent of QR, i.e. all straight lines converge on the same point on the abscissa. The equation (of the kind $y = cost - b\ x$) indicates that the intercept on the ordinate: $P_{i,O_2} \times QR/\left[1 - F_{i,O_2}(1 - QR)\right]$ and the slope: $QR/\left[1 - F_{i,O_2}(1 - QR)\right]$ are greater the greater QR. Consider point A in Fig. 3.48: the same respiratory condition (i.e. the same P_{alv,CO_2} and P_{alv,O_2}) is attained at 5000 and 7000 m if $QR = 0.6$ at 5000 m and 1.4 at 7000 m, i.e. the *resistance to altitude is greater the greater the QR*. This shows that glucides ($QR = 1$) are more suitable than fat ($QR = 0.7$) to resist hypoxia at high altitudes. Since the maximum 'internal' QR is 1 (see Table at the end of Sect. 3.6.2), an 'external' $QR = 1.4$ is attainable only by hyperventilation, i.e. with a pulmonary ventilation greater than the metabolic requirements; this can be attained by depleting the CO_2 reserves of our organism, a condition that cannot be maintained endlessly. Another case in which $QR > 1$ is the transformation of glucides into lipids, but also this is a necessarily transient process.

The curve *Norm.* in Fig. 3.48 shows the respiratory condition of a non-acclimatized subject increasing his altitude while breathing air; it can be seen that his alveolar

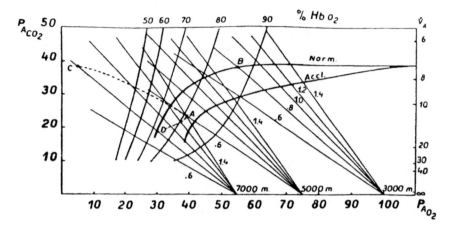

Fig. 3.48 Same abscissae and ordinates as in Fig. 3.47. The straight lines depart from three values of $P_{alv,O_2} = P_{i,O_2}$ on the abscissa corresponding to the altitudes of 3000, 5000 and 7000 m above sea level. The different slopes in each set of lines refer to different values of the respiratory quotient as indicated. The curves *Norm* and *Accl* indicate the respiratory conditions of two subjects non-acclimatized respectively acclimatized, to low barometric pressure (from Margaria R and De Caro L, *Principi di Fisiologia Umana*, Vallardi, Milano, 1977)

ventilation $\dot{V}_{alv}$ (ordinate on the right) remains about constant until P_{alv,O_2} (bottom abscissa) falls to about 50–60 mm Hg; at this point (B) the carotid and aortic bodies increase his ventilation until a maximum altitude of 7000 m is attained with a $QR = 0.8$ and $P_{alv,CO_2} = 20$ mmHg (the same condition as at 14,500 m when breathing pure oxygen Fig. 3.47). The point C indicate an hypothetical condition in which the carotid and aortic bodies would not respond to hypoxia maintaining unaltered $\dot{V}_{alv}$ and the $QR = 0.8$ with increasing altitude; in this case the P_{alv,O_2} would fall to about 5 mmHg at 7000 m incompatible with life. To avoid this condition $\dot{V}_{alv}$ increases attaining point A with a confortable $P_{alv,O_2} = 40$ mmHg and $P_{alv,CO_2} \sim 20$ mmHg, but with a $QR = 1.4$ too high to be sustained at equilibrium; the interrupted line A to D indicates that a correct $QR = 0.8$ is finally attained with a decrease of P_{alv,O_2} and a further increase in $\dot{V}_{alv}$. The curve *Accl.* refers to a subject acclimatized to high altitude: differently from the *Norm* subject, its alveolar ventilation starts to increase at lower partial pressure of oxygen, i.e. at lower altitude due to a greater sensitivity of his carotid and aortic bodies and attains a final conditions at 7000 m with a higher P_{alv,O_2} (~40 mmHg).

In conclusion: the respiratory condition of a subject breathing air at high altitude depends from: (i) the alveolar ventilation $\dot{V}_{alv}$ that must be increased; (ii) the respiratory quotient QR: the resistance to altitude is greater the greater QR; (iii) the resistance to alkalosis: introducing acidifying substances may help, e.g. ammonium chloride ($NH_4^+ + Cl^- \rightarrow NH_3 + H^+ + Cl^-$); (iv) the capabilities of the kidneys to eliminate bicarbonate ions HCO_3^- (pH $= 6.1 + \log [HCO_3^-]/\alpha\ pCO_2$).

3.11 Diving

During underwater immersion the pressure P increases of 1 Atm every 10 m corresponding to 1.033 kg/cm^2 every 10 m (1 l of water weights 1 kg, occupies a cube of 10 cm × 10 cm × 10 cm = 1000 cm^3 and leans over the base of 100 cm^2 resulting in a pressure of 1 kg/100 cm^2, 10 cubes one over the other will have a height of 1 m resulting in a pressure of 10 kg/100 cm^2 and 100 cubes one over the other will have a height of 10 m resulting in a pressure of 100 kg/100 cm^2, i.e. 1 kg/cm^2).

3.11.1 Diving in Apnea

We have to consider three laws: (i) $P + dgh =$ constant (see Sect. 1.3.1); (ii) $P \times V$ = constant (Boyle's law, see Sect. 3.5.2); (iii) the *Archimedes' principle* stating that the upward buoyant force that is exerted on a body immersed in a fluid is equal to the weight of the fluid that the body displaces and acts in the upward direction at the center of mass of the displaced fluid. According to the Archimede's principle the net force F acting vertically on the immersed body is:

$F = m_{body} \, g - m_{liquid} \, g = \delta_{body} \, V_{body} \, g - \delta_{liquid} \, V_{liquid} \, g$ and since $V_{body} = V_{liquid}$, it follows that $F = V_{body} \, g \, (\delta_{body} - \delta_{liquid}) = V_{body} \, \delta_{body} \, g \, (1 - \delta_{liquid}/\delta_{body})$, i.e.:

$F = m_{body} \, g \, (1 - \delta_{liquid}/\delta_{body})$. It follows that the net force F acting vertically on the immersed body will be nil when $\delta_{liquid} = \delta_{body}$, will be directed upwards when $\delta_{liquid} > \delta_{body}$ and downwards when $\delta_{liquid} < \delta_{body}$. When swimming deeper in apnea the volume of the lungs decreases due to the increase in pressure (Boyle's law) with a consequence that the *average* density of the body increases; near the surface one must swim actively towards the bottom, for example to get a fish, but beyond a certain depth the body flows freely downwards towards the prey! However there is a critical point to consider: under water, suppose 10 m, both P_{alv,O_2} and P_{alv,CO_2} are kept high by the increased water pressure; when a limit is attained due to the effect of the increased P_{alv,CO_2} on the respiratory center, the subject increases his upward lift; by doing so however the outside pressure and with it the P_{alv,O_2} may decrease to a limit incompatible with his consciousness: in this case the subject loses consciousness and under the stimulus of the P_{alv,CO_2} inhale water and die. Another problem of deep diving in apnea is represented by the reduction of the pulmonary volume with depth below that of the residual air, i.e. below the volume we can voluntarily decrease with a maximal expiration (Fig. 3.10). At this volume the structures surrounding the lungs (chest wall and diaphragm) become relatively rigid resisting a further decrease in volume. Diving further below this point results in a pressure in the lungs smaller than that of blood, in equilibrium with the outside pressure, with the consequence that the alveolar capillaries will outstretch and hemorrhages may ensue. This problem, observed in humans, poses the question of how marine creatures provided by lungs such as dolphins and whales can reach so great depths under sea level (~1500 m!). The experiment described in Fig. 3.49 was devised to answer this question. A dolphin

Fig. 3.49 The experimental setup for deep diving experiments in the open ocean. The porpoise dives down when the go signal is turned on. He pushes the plunger on the end of the diving test switch and then returns to exhale into the funnel before surfacing (right). Picture below was taken at a depth of 300 m when Tuffy pushed the switch: note the thoracic collapse behind the left flipper (from Ridgway SH, Scronce BL and Kanwisher J *Science* 166: 1651–1654, 1969)

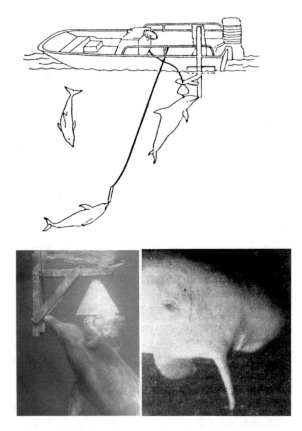

(Tuffy) was instructed to swim just below sea level in one case, and for the same amount of time at 300 ms below sea level in another case. The height attained under sea level was confirmed by Tuffy pushing a switch. After both swims Tuffy exhaled in a cone (Fig. 3.49). This experiments showed that the oxygen consumption was greater when Tuffy swim on the surface than on depth. This can be explained by assuming that Tuffy's lungs are *completely atelectatic* at 300 m so that no gas exchange takes place between alveolar air and blood. This hypothesis is qualitatively consistent with the evident cavity that one can observe in Fig. 3.49 on Tuffy's body when Tuffy photographed himself by pushing a switch at depth. The 7 l of Tuffy's lungs on the surface reduce to 7000 ml $\times$ 1 Atm $= x \times$ 30 Atm, i.e. to 7000/30 = 233 ml, which is just the volume of the rigid bone cavities near the nose: at depth, air is compressed there as in a spray can re-expanding the lungs during the ascent.

3.11.2 Diving with Respirator

An old abandoned respirator is the *oxygen breathing respirator* (closed circuit); its circuit is similar to that schematized in Fig. 3.17: instead of the bell, a sac *containing pure oxygen* is bound to the chest, a three ways valve group, a CO_2 absorber, a nose plug and a gas bottle of oxygen to refill the sac when O_2 is progressively consumed. *It is important that the sac and the lungs contain pure oxygen without nitrogen*: for this reason the sac-lung system must be emptied and refilled with oxygen a few times before immersion and care must be taken not to inspire air after this procedure. In fact if air enter the system, the CO_2 stimulus being absent due to the CO_2 absorber, the subject will lose consciousness due to hypoxia while breathing nitrogen instead of oxygen from and into the sac. Another problem is the toxic effect of oxygen at high pressures: at 3–4 atmospheres of pure oxygen the subject loses consciousness and may die. The advantages of the pure oxygen respirator is that no bubbles appear on the surface (an advantage in war) and to avoid embolism (see below).

Most common is the *air respirator* (open circuit) consisting of two compressed air tanks carried on the back of the subject and a pressure-reducing valve system on the mouth of the subject allowing inspiring air *at the water pressure* and expiring air into the water (bubbles of air are visible on the water surface). This respirator may cause *narcosis of nitrogen* at very high pressures (~30–40 Atm), but more commonly may cause *embolism* as described below even at lower depths.

The partial pressure of nitrogen and oxygen in the air inspired from the tanks increases with depth in the lungs (not that of CO_2, and we will see why later). According to Henry's law (see Sect. 3.9) the concentration of gas dissolved into a liquid increases in proportion to its partial pressure. The inert nitrogen gas (79% in air) takes some time to dissolve into the alveolar surface, to travel in the alveolar capillary blood, in the main blood vessels, in the tissues capillaries and in the cells of the whole body, particularly those with a high concentration of lipids, such as those of the brain, where it is more soluble. If enough time is allowed to complete this process, tissues will saturate with nitrogen: this is not the case when deep diving in apnea where the full immersion time is necessarily limited. When tissues are saturated with nitrogen at a given depth, enough time must be allowed during the ascent to perform its backward journey, from its dissolved condition into the cells to the alveolar air: this time, i.e. the duration of the *decompression* during the ascent to the surface, must be adequately long. In case of a rush to the surface, the pressure keeping the nitrogen molecules dissolved in the tissues falls abruptly and (as when opening a bottle of a gaseous drink) nitrogen sets free in bubbles *within* the cells and by doing so divaricates layers of tissue causing a damage particularly evident into the brain. This condition, called *embolism*, is made particularly severe by motion (as shacking a bottle with a liquid saturated with gas). Bubble producing by motion is called *cavitation*. If embolism occurs the first thing to do is to take back the subject at his initial depth; in case this is not possible, to bring the subject in a hyperbaric chamber as soon as possible. To avoid embolism the subject must rise to the surface with a time trend depending on: (i) the reduction of pressure per unit of vertical distance during the lift; (ii) the velocity of nitrogen leaving the tissues. As it will be described below both mechanisms work in the same way *indicating to reduce progressively the velocity of the ascent.*

(i) Since $PV =$ constant the volume V of an eventual gas bubble will double if P is halved: this means that the possibility to get embolism would equal in the lifts from 4 atmospheres (30 m depth), to 2 atmospheres (10 m depth) and from 10 m to the surface, i.e. the velocity of the lift must decrease exponentially when approaching the surface.

(ii) The velocity of nitrogen leaving the tissues depends on nitrogen molecules concentration $[N_2]$, i.e. $-dN_2/dt = k[N_2]$; this is called a *first order* velocity of reaction (the minus sign indicates that the concentration of N_2 *decreases* with time). To understand the meaning of this equation, whose trend describes several processes in biology, imagine dropping an effervescent tablet in a glass of water; at the beginning, when the tablet surface is constant, the velocity of gas production is constant (i.e. $dx/dt =$ constant): this is called a *zero order* velocity of reaction. However when the surface x of the tablet decreases the velocity of bubbles production progressively decreases, i.e. $-dx/dt = kx$: a *first order* reaction. Another example: the reaction of an enzyme in excess of substrate will be a *zero order* process, but as soon as the concentration of substrate decreases below that of the enzyme the velocity becomes that of a *first order* process.

In order to know the concentration of nitrogen N_2 as a function of time t we must remember (or to accept for those who did not follow a math course) that $\int dy/y = lny + const$. In our case: $-dN_2/dt = kN_2$ can be rewritten as $\int dt = -1/k \int dN_2/N_2$ and $t = -1/k \ln N_2 + const$. When $t = 0$, i.e. at the beginning of the lift, $const = 1/k \ln N_2O$ and substituting the value of the constant: $t = -1/k \ln N_2 + 1/k \ln N_2O$, which can be written as $t = (-1/k) \times (\ln N_2 - \ln N_2O) = (-1/k) \times \ln N_2/N_2O$. We can now calculate the *time of semi reaction*: $t_{1/2} = 2.3 \times 0.3/k = 0.69/k$, *which holds for any first order natural process* (remember that $2.3 \times \ln = \log$ and that $\log \frac{1}{2} = -0.3$).

The time course of nitrogen liberation, as well as that of *any* first order natural processes will then be $\ln N_2 = \ln N_2O - kt$, a straight line if expressed in a logarithmic form, or as an exponential decay in a non logarithmic form:

$$N_2 = e^{(\ln N_2O - kt)} = e^{(\ln N_2O)} \times e^{-kt}, \text{ i.e. } N_2 = N_2O \times e^{-kt}.$$

3.12 Pneumothorax

The pneumothorax is the presence of air in the pleural cavity: the lung collapses plus or minus depending on the volume of air contained between visceral and parietal pleurae. The pressure P_L of the air contained between chest wall and lung is less than the atmospheric pressure P_B due to the residual tendency of the lung to collapse

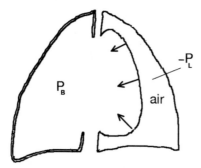

Fig. 3.50 The lung on the left of the Figure shows the normal condition when the two pleurae, parietal and visceral, are in contact, whereas the lung on the right is partially collapsed because of air into the pleural space: the *pneumothorax*. Due to the residual tendency of the lungs to collapse (arrows), the pressure of the air in the pneumothorax P_L is less than barometric pressure P_B

(Fig. 3.50). There are several kinds of pneumothorax: (i) the *clinical* or *closed* pneumothorax, which is made on purpose by introducing air between the two pleurae with a needle in order to allow the edges of a pulmonary wound to approach favoring healing; (ii) the *open* pneumothorax due to an external wound, which is dangerous because of possible infections and cooling due to evaporation of the pleural liquid, or to an internal wound on the surface of the lung opening through the visceral pleura into the pleural cavity; (iii) the most dangerous, the *valvular* pneumothorax: this occurs when as a consequence of an external wound or of an internal cavity on the surface of the lung a flap of tissue allowing enter of air into the pneumothorax during inspiration and closes the output during expiration with the result that air is 'pumped' within the thorax, pushing against the mediastinum and the heart.

3.12.1 Precautions to Take in Case of a Patient with Pneumothorax

Climbing at high altitude the atmospheric pressure P_B decreases (Fig. 3.3). Since $PV = $ constant, a reduction of P_B causes an increase in the volume of air in a closed pneumothorax: air would expand pushing against the mediastinum and the heart. For this reason, before moving a patient with pneumothorax towards a higher altitude location, it is necessary to withdraw the correct volume of air from the pneumothorax. Furthermore: the air of the pneumothorax is progressively reabsorbed by blood, as in all closed cavities of our body. If the cavity has collapsible walls, the cavity disappears, if the walls are rigid, as for example the nose sinuses, the cavity will fill with liquid. Figure 3.51 explains why.

As shown in Fig. 3.51, for a *given volume of oxygen consumed and carbon dioxide produced* (e.g. 5 ml% with a $QR = 1$) the O_2 pressure decreases of about 50 mmHg

Fig. 3.51 Curves of dissociation of blood for (volume of gas in 100 ml of blood) oxygen O_2tot and carbon dioxide CO_2tot with their components of O_2 and CO_2 physically dissolved according to Henry's law (straight lines at the bottom). The rectangles show the large fall in pressure in O_2 and the small increase in pressure of CO_2 when blood flows from the arterial to the venous extremity of the capillary (from Henderson LJ *Blood*, Yale University Press, 1928)

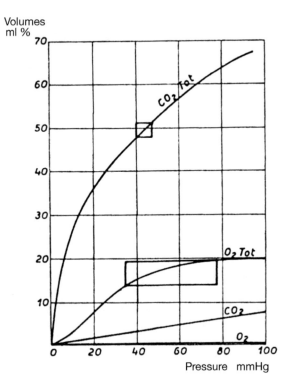

whereas that of CO_2 increases of about 5 mmHg only; this is due to the fact that the affinity of blood for CO_2 is greater than that for O_2 (as we will see in detail later). It follows that during oxygen consumption, the *venous extremity* of the capillary behaves as a pump withdrawing all gases from tissues in equilibrium with the atmospheric pressure, because *the large fall in partial pressure of oxygen consumed implies an increase in partial pressures of* CO_2 *and nitrogen* since the total pressure is maintained equal to atmospheric, as a consequence all air in the pneumothorax is reabsorbed. In conclusion: the two precautions to take with a closed pneumothorax is withdrawing air before ascent to higher altitude and to add air when the volume initially injected is reduced for the reason explained above.

3.12.2 Respiratory Mechanics with the Pneumothorax

In normal condition the volume changes of the lung exactly equal those of the thorax wall because the two surfaces adhere closely with a small amount of incompressible interstitial liquid between them, i.e. $\Delta V_{thorax} = \Delta V_{lungs}$. When coughing the acceleration of the lungs equals that of the thorax wall, a condition that may be

Fig. 3.52 Schematic
representation of the
expansion E (ordinate) of the
thorax (T) and of the lung (L)
as a function of time (t) in
the closed pneumothorax. (a)
Amplitude of the lung
volume changes. (b)
Amplitude of the thorax
volume changes (from
Margaria R and De Caro L
Principi di fisiologia umana,
Vallardi, Milano, 1977)

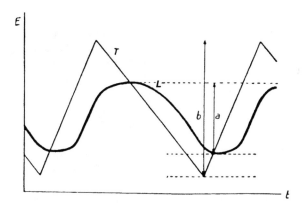

dangerous in case of a lesion of the lung. On the contrary in the pneumothorax
$\Delta V_{thorax} = \Delta V_{lung} + \Delta V_{air}$ (Fig. 3.52).

The mechanics of breathing with a pneumothorax is characterized by: (i) a delay
between thorax and lung movement; this is due to the fact that motion of thorax wall
causes first compression, or expansion, of the pneumothorax air, which subsequently
acts on lung surface; (ii) the tidal volume of the lung ΔV_{lung} is less than the amplitude
of the thorax volume changes ΔV_{thorax}: this is because $\Delta V_{thorax} = \Delta V_{lung} + \Delta V_{air}$
and ΔV_{air} increases during inspiration due to the increased tendency of lung to
collapse (vice versa during expiration) (Fig. 3.50).

3.13 Ventilation-Perfusion Ratio

The exchanges of the respiratory gasses between blood and air require both alve-
olar ventilation $\dot{V}_A$ and blood perfusion $\dot{Q}$, i.e. an adequate $\dot{V}_A/\dot{Q}$, ratio. To avoid
losses of alveolar ventilation $\dot{V}_A$ and of blood perfusion $\dot{Q}$, the optimal $\dot{V}_A/\dot{Q}$, ratio
should equal unity. Let's consider the two possible extremes of the $\dot{V}_A/\dot{Q}$, ratio.
The first extreme is the *physiological dead space*; up to now we have considered the
250 ml of the *anatomical dead space*, i.e. the volume of air surrounded by thick walls
(nose, trachea and bronchi) impermeable to gasses. However the *total physiological
dead space* equals *anatomical dead space* + *alveolar dead space*; to understand
the alveolar dead space we can imagine an alveolus so big and a flow of air so fast
that no complete equilibrium is attained between hemoglobin and air at its center.
One extreme of the physiological dead space is $\dot{V}_A/\dot{Q} = \infty$, i.e. unused ventilation. The other extreme is the *shunt*, when $\dot{V}_A/\dot{Q}, = 0$, i.e. when we have perfusion
without ventilation (Fig. 3.53). The result is partially reduced hemoglobin when the
oxygenated hemoglobin mixes with the reduced one.

The partial pressures of oxygen and carbon dioxide at the two extremes of the
$\dot{V}_A/\dot{Q}$ ratio are shown in Fig. 3.54. On the left the gas pressures are those of the

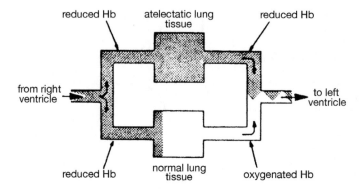

Fig. 3.53 The blood with reduced hemoglobin due to an atelectatic lung zone mixes with oxygenated hemoglobin resulting in only partially oxygenated arterial blood (anoxic anoxia). This kind of anoxia cannot be compensated by hyperventilation because the blood passing through the normal lung tissue is already saturated with oxygen (modified from Green JH *An Introduction to Human Physiology*, Oxford University Press, London, 1968)

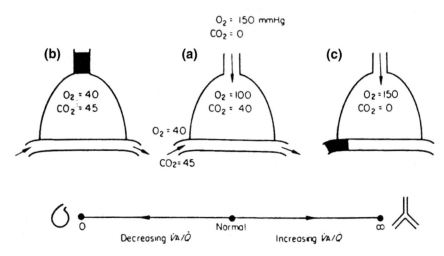

Fig. 3.54 Extremes of ventilation-perfusion ratio. Normal gas exchange is seen in (**a**) where the balance of ventilation and blood flow is such that the alveolar $P_{alv,O_2} = 100$ mmHg and the $P_{alv,CO_2} = 40$ mmHg. In (**b**), ventilation has been completely obstructed, the ventilation-perfusion ratio is nil, and the alveolar gas tensions are those of mixed venous blood (the symbol shows the right ventricle). In (**c**), blood flow has been stopped, the ventilation-perfusion ratio is infinitely high and the alveolar gas tensions are those of inspired gas (the symbol shows the trachea). The line at the bottom shows how the ventilation-perfusion ratio changes between these two extremes (from West JB *Respiratory physiology*, Lippincott Williams & Wilkins, 2004)

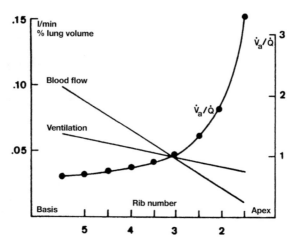

Fig. 3.55 Distribution of ventilation and blood flow in an erect lung. The left ordinate and the straight lines show the *regional* flow of blood and air. The ordinate on the right and the filled points show the ventilation-perfusion ratio at each level. Note that the ventilation-perfusion ratio decreases from the apex to the basis of the lung (from West JB *Ventilation/Blood flow and Gas Exchange*, Oxford, Blackwell, 1970)

venous blood in the right ventricle and remain unchanged when venous blood flows through the lungs because airflow trough the lungs is impeded by obstruction of the trachea, i.e. $\dot{V}_A/\dot{Q} = 0$. The opposite condition is shown on the right hand of the figure, where blood flow through the lungs is impeded by closure of the pulmonary artery, i.e. $\dot{V}_A/\dot{Q} = \infty$; in this case the gas pressures in the lungs equal those of the ambient air. In the center of the Figure is illustrated the normal condition when both flows are allowed: that of venous blood and that from the ambient air, i.e. $\dot{V}_A/\dot{Q} = 1$, resulting in the normal $P_{alv,O_2} = 100$ mmHg and $P_{alv,CO_2} = 40$ mmHg in the arterial blood.

The left ordinate and the straight lines of Fig. 3.55 show the *regional* flow of blood and air (i.e. the ratio between total flow and % volume of a strip of lung) from bottom to apex of a lung in an erect position. The ordinate on the right shows the ventilation-perfusion ratio at each level. The diaphragm, the principal inspiratory muscle, and gravity pull together in making ventilation and blood flow greater at the basis of the lung.

However blood flow decreases with height more quickly than ventilation; the two flows equal at the level of the third rib. Above this level $\dot{V}_A/\dot{Q}$ increases exponentially approaching the value of the dead space at the apex of the lung (*c*, in Fig. 3.54); below this level $\dot{V}_A/\dot{Q}$, decreases more slowly towards the shunt condition (*b*, in Fig. 3.54).

The curve in Fig. 3.56 *indicates all possible compositions of alveolar air pressure in a lung connected on one side to an air ambient having gas composition I and on the other side to venous blood with gas composition $\overline{V}$.* What determines its shape? As we have seen in Fig. 3.48 a fan of iso-QR straight lines departs in the air phase with slope 45° when $QR = CO_2/O_2 = 1$, higher slope when $QR > 1$ and lower slope when $QR < 1$ according to the *alveolar air equation*. Similarly, we can imagine a fan of iso-QR lines departing from the $\overline{V}$ point in the graph of Fig. 3.56 for the liquid (blood) phase. *The iso-QR lines departing from the $\overline{V}$ point reflect the difference between the two saturation curves of blood for CO_2 and O_2: for a given volume*

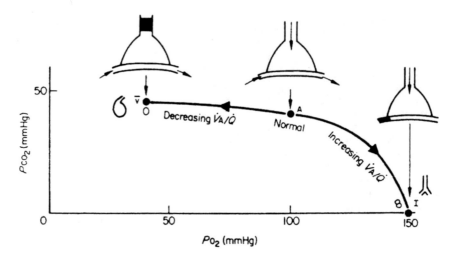

Fig. 3.56 The oxygen-carbon dioxide diagram. The composition of the air in the three lung units of Fig. 3.54 is shown. The line joining them is known as the '*ventilation-perfusion ratio line*' because it represents all the possible alveolar gas compositions as the ventilation-perfusion ratio decreases to nil, or increases to infinity (from West JB *Respiratory physiology*, Lippincott Williams & Wilkins, 2004)

of CO_2 eliminated, the partial pressure of CO_2 falls much less than the increase in partial pressure of O_2 for the same volume of oxygen absorbed (i.e. for the same QR). It follows that in the iso-QR lines departing from the $\overline{V}$ point the vertical fall of pressure on the ordinate (CO_2 eliminated) is much less than the horizontal increase of pressure on the abscissa (O_2 absorbed). The points where the iso-QR lines departing from the I point on the abscissa match those departing from the $\overline{V}$ point on the ordinate define the shape of the curve in Figs. 3.56, 3.57 and 3.58.

Figure 3.58 shows schematically what explained above: the match of the iso-QR lines departing from the venous blood $\overline{V}$ and from the air I defines the trend of the equilibrium point between air and blood; *the thick line indicate the points where the partial pressures in the liquid phase (blood) equal those in the alveolar air*. Actually a perfect equilibrium is not completely attained at the point of intersection between blood (curves) and air (straight lines) because the partial pressure of oxygen in the blood, deriving from mixing different zones of the lung, is less (~4 mmHg) than that in the alveolar air. This is because, as shown in Fig. 3.51, hemoglobin at the apex of the lung saturates at a pO_2 of only about 100 mmHg resulting in a mixed average pO_2 pressure in the blood (curves) lower than that in the gaseous environment predicted by the alveolar air equation (straight lines).

During muscular exercise the *ventilation/perfusion* ratio improves because with increasing cardiac output the difference between blood flow at the apex and the basis of the lungs decreases. If, on the contrary, alveolar ventilation decreases relative to the demand, the increase in pCO_2 causes bronchial dilatation and the decrease in

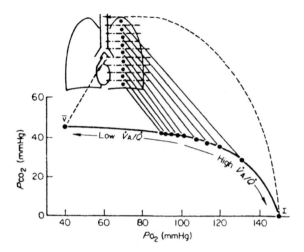

Fig. 3.57 Oxygen-carbon dioxide diagram showing how the ventilation-perfusion ratio $\dot{V}_A/\dot{Q}$, increasing from bottom to top of the lung, determines the regional composition of alveolar gas. The lung is divided into nine imaginary horizontal slices each of which has its own position on the ventilation-perfusion ratio line (Fig. 3.55). It can be seen that P_{alv,O_2} will increase up the lung in so far as the points move horizontally to the right, and the P_{alv,CO_2} will fall as the points move vertically downwards. Dashed lines show the composition of mixed venous (pulmonary artery) blood and inspired (tracheal) air (from West JB *Respiratory physiology*, Lippincott Williams & Wilkins, 2004)

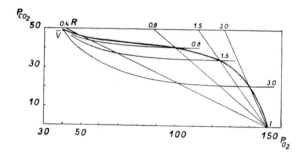

Fig. 3.58 The curves departing at the left from the CO_2 *and* O_2 *pressure values in the mixed venous blood* $\overline{V}$ show how these values change *in the liquid blood phase* when blood equilibrates with the alveolar air by losing CO_2 and acquiring O_2 as indicated by the respiratory quotient QR = $CO_2/O_2 = 0.4$–3.0. The straight lines departing at the right from the CO_2 *and* O_2 *pressures values in the inspired ambient air* I show how these pressures change *in the alveolar air* (see Fig. 3.48) when CO_2 is acquired and O_2 is lost as indicated by the same respiratory quotient QR = CO_2/O_2 = 0.4–3.0. In both cases each line refers to a given ratio QR between volume of CO_2 lost and O_2 gained. It follows that the continuous thick line joining the intersection points of *blood* curves and *air* straight lines represents the unique pCO_2–pO_2 possible combinations in the alveolar air after blood-air equilibrium: no points are possible outside this line which represent the perfect equilibrium between air and blood (modified from Margaria R and De Caro L *Principi di fisiologia umana*, Vallardi, Milano, 1977)

pO_2 causes *in the lungs* vasoconstriction tending to maintain constant the ventilation/perfusion ratio.

3.14 Gas Transport in Blood

Something of what follows has already mentioned previously (Fig. 3.51). The *respiratory* gas transport is studied by constructing the *dissociation curves* of blood for O_2 and CO_2. These curves have on the ordinate the ml of gas found in the blood when the blood is in equilibrium with the pressure of the gas plotted on the abscissa (*back pressure*, i.e. the pressure that must be exerted to maintain the gas attached to the blood against its tendency to dissociate from the blood). Figure 3.59 shows a *tonometer* used to equilibrate a small amount of blood made incoagulable with a large volume of gaseous mixture of known composition.

3.14.1 Oxygen

Contrary to CO_2, as we will see, the oxygen O_2 can be transported by blood in only two ways: physically dissolved (Henry's law) and combined with the hemoglobin. The volume of oxygen physically dissolved increases linearly with pressure attaining 0.3 ml O_2/100 ml *of blood at* $pO_2 = 100$ mmHg. The volume of oxygen bound to hemoglobin attains *a maximum of* 20 ml/100 ml *of blood at* $pO_2 = 100$ mmHg; this is in contrast with CO_2 whose concentration in the blood continues to increase with increasing pCO_2 (Fig. 3.51). The *dissociation curve* of blood for oxygen, i.e. the relationship between HbO_2 and pO_2 (Fig. 3.61), has two ordinates: (i) ml O_2/ml of blood and (ii) *% saturation* of hemoglobin with oxygen: [HbO_2/(Hb + HbO_2)] $\times$ 100. The *concentration of Hb in blood is* 15 g/100 ml, resulting in a maximal saturation of 20 ml O_2/15 g of Hb = 1.33 ml O_2/g of Hb. As known, one mole of gas

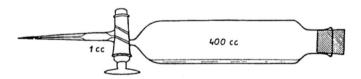

Fig. 3.59 Tonometer used to equilibrate a small sample of incoagulable blood with 400 ml of a gaseous mixture of known composition; equilibrium is attained by slow rotation of the tonometer at a given temperature resulting in a very thin layer of blood on its walls. The tonometer is provided at one extremity with a pipette of 1 ml containing the blood equilibrated with the gas mixture to decant it directly into the Van Slike apparatus, which measures the volume of the gasses by extracting them from the blood with a vacuum pump. The operation is repeated with different gas mixtures to determine the gas concentration in blood as a function of the gas pressure (Fig. 3.51) (from Margaria R and De Caro L *Principi di fisiologia umana*, Vallardi, Milano, 1977)

Fig. 3.60 a Dissociation curve of human blood at 38 °C, pH = 7.4. **b** Oxygen dissociation curve of myoglobin under similar conditions (from Roughton FJW *Handbook of Respiratory Physiology*, U.S.A.A.F. Aviation School of Medicine, 1954)

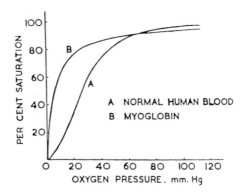

occupies 22.4 l at STP (1 Atm and 0 °C), it follows that the grams of hemoglobin bound to one mole of oxygen are: 1.33 ml O_2: 1 g Hb = 22,400 ml: x g Hb, i.e. x g Hb = 22,400/1.33 ~ 17,000 g Hb bind one mole of oxygen. The molecular weight of Hb is 68,000, this means that one mole of Hb will bind four moles of oxygen, i.e. one molecule of Hb binds 4 molecules of oxygen (since in both cases one mole contains 6×10^{23} molecules).

The percentage of iron, F_e^{++} (atomic weight 56) in Hb is 0.33%; it follows that: one mole of iron, i.e. 56 g F_e^{++}: x g Hb = 0.33 g F_e^{++}: 100 g Hb, i.e. x = 5600/0.33 ~ 17,000. In conclusion, one molecule of HbO_2 contains four atoms of iron, each bound to one molecule of oxygen.

The molecule of *myoglobin* has molecular weight ~17,000, one quarter that of hemoglobin and binds one molecule of oxygen. The dissociation curve of Mb can be determined as follows: $Mb + O_2 \rightleftarrows MbO_2$, the velocity of association with oxygen will be $v_1 = k_1 [Mb] \times [O_2]$, that of dissociation $v_2 = k_2 [MbO_2]$; at the equilibrium $v_1 = v_2$, i.e. $k_1[Mb] \times [O_2] = k_2 [MbO_2]$. The ratio $k_1/k_2 = K$, *the constant of equilibrium*, is $K = [MbO_2]/([Mb] \times [O_2])$, i.e. $[MbO_2] = K \times [Mb] \times [O_2]$. The *saturation* on the ordinate of Fig. 3.60 is $y/100 = [MbO_2]/([Mb] + [MbO_2])$. Since $[MbO_2] = K \times [Mb] \times [O_2]$ we can substitute: $y/100 = (K \times [Mb] \times [O_2])/([Mb] + (K \times [Mb] \times [O_2]) = K \times [O_2]/(1 + K \times [O_2])$. In Fig. 3.60, the % saturation is plotted as a function of oxygen pressure instead of oxygen concentration; i.e. $y'/100 = K'pO_2/(1 + K'pO_2)$. From this equation it can be seen that when the pO_2 tends to infinity the ratio tends to one, i.e. saturation tends to 100%.

The hyperbolic trend of myoglobin saturation as a function of pO_2, differs from that of hemoglobin, which is made up by four molecules similar to myoglobin. To define the saturation curve of the molecule of hemoglobin the procedure described above must be repeated four times, with four constants of equilibrium K_1, K_2, K_3, K_4 of O_2 attachments to the four F_e^{++} of the molecule of the hemoglobin. The last constant K_4 is about 100 times greater than the first three, which are similar, giving to the curve of dissociation of hemoglobin its characteristic shape: slow initial slope followed by a sharp increase (Adair equation).

Why four units in the hemoglobin molecule? The dissociation curves in Figs. 3.51 and 3.60 are characterized by: (i) their slope, i.e. by the ratio between increment (or decrement) of the ordinate and the corresponding change on the abscissa, which represents the *affinity* of the molecule for O_2 or CO_2; and (ii) the height of the curve, which represents the *capacity*, i.e. the amount of gas (O_2 or CO_2) that can be transported. The affinity indicates the volume of gas delivered or acquired for a given change of its partial pressure. From Fig. 3.60 it can be seen that the affinity of myoglobin for oxygen is maximal at $pO_2 = 0$, whereas that of hemoglobin is maximal when $pO_2 = 30$–40 mmHg and minimal at high pO_2 pressure; this is consistent with Figs. 3.47 and 3.48 showing that the alveolar ventilation (right hand ordinate) changes little at high pO_2 pressure whereas increases in the range of alveolar pressures (30–40 mmHg) just when the affinity for hemoglobin is maximal, i.e. small changes in pO_2 cause relevant oxygen exchanges between air, tissues and blood. This is in contrast with the curve of myoglobin showing that oxygen is almost totally retained at 30–40 mmHg, i.e. in the pressure range where oxygen exchange takes place at the tissues level. This is the advantage to have four oxygen binding structures instead of one in the Hb molecule.

As shown in Fig. 3.61 CO_2 decreases the affinity of hemoglobin for O_2. As described below this is due to both: (i) the molecule of CO_2 as such and (ii) the acidification of blood [H^+]:

$$CO_2 + H_2O \leftrightarrows H_2CO_3 \leftrightarrows H^+ + HCO_3^- \leftrightarrows H^+ + CO_3^{--}$$

The CO_2 combines with water with a reaction facilitated by the *enzyme carbonic anhydrase* forming carbonic acid, which dissociates into hydrogen ion H^+ and the bicarbonate ion HCO_3^-, which, in turn, dissociates into hydrogen ion and the carbonate ion CO_3^{--}. Therefore CO_2 increases the concentration of H^+, which decreases the affinity of hemoglobin for oxygen. This is called Bohr's effect, which will be treated afterwards.

The experiment described in Fig. 3.62 shows that acidification is not the only factor causing a decrease of the affinity of hemoglobin for oxygen.

Since $Na^+ + HCO_3^- \leftrightarrows Na^+ + OH^- + CO_2$ the reduced affinity of hemoglobin for oxygen, indicated by displacement of the curve to the right, must be due to [CO_2] itself independent of [H^+], which in the experiment of Fig. 3.62 is maintained constant. We can therefore write: $HbO_2 + CO_2 \leftrightarrows HbCO_2 + O_2$, keeping however well in mind that, contrary to CO, which binds to the same site of oxygen (the F_e^{++} ions), CO_2 binds to amino groups in the Hb molecule: $-NH_2 + CO_2 \rightleftarrows NHCOOH$.

Let's now consider the effect of [H^+] only (Bohr's effect), leaving aside the CO_2, i.e. $HbH \leftrightarrows Hb^- + H^+$. The dissociation velocity is $v_1 = k_1[HbH]$, that of association is $v_2 = k_2[Hb^-] \times [H^+]$. At equilibrium $v_1 = v_2$, i.e. $k_1[HbH] = k_2[Hb^-] \times [H^+]$ and $k_1/k_2 = K = ([Hb^-] \times [H^+])/[HbH]$; $1/[H^+] = 1/K \times ([Hb^-]/[HbH])$. In logarithmic form $\log 1/[H^+] = \log 1/K + \log([Hb^-]/[HbH])$; i.e. $pH = pK + \log([Hb^-]/[HbH])$, from which it can be seen that the greater the pH the more dissociate is Hb, which is

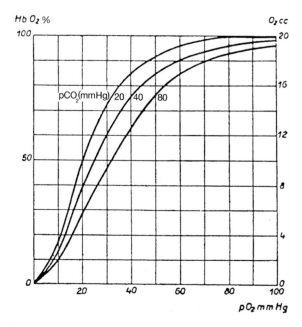

Fig. 3.61 Dissociations curves of blood for oxygen. On the left hand ordinate is given the % *saturation* of Hb with oxygen, i.e. [HbO₂/(Hb + HbO₂)] × 100. On the right hand ordinate are given the ml of O₂ bound to Hb in 100 ml of blood. On the abscissa is given the partial pressure of O₂ in mmHg. The three curves refer to the same blood in equilibrium with three different solutions with partial pressure of CO₂ of 20, 40 and 80 mmHg; the respective values of pH being 7.45, 7.44 and 7.24. The central curve with $pCO_2 = 40$ mmHg refer to the normal physiological condition (see Fig. 3.54a) (from Margaria R and De Caro L *Principi di fisiologia umana*, Vallardi, Milano, 1977)

a weak acid. The oxygenation of hemoglobin makes it a stronger acid: $HbH + O_2 \leftrightarrows HbO_2^- + H^+$, i.e. HbO_2^-, has a greater tendency to release H^+ than Hb (Fig. 3.63).

The curves in Fig. 3.63 are obtained by *titration* of hemoglobin HbH, which is a weak acid (a strong acid is always completely dissociated), by adding Na^+OH^-: $HbH + Na^+ + OH^- \leftrightarrows Hb^- + Na^+ + H_2O$, and $HbO_2H + Na^+ + OH^- \leftrightarrows HbO_2^- + Na^+ + H_2O$. The two curves so obtained in Fig. 3.63 run from dissociation equal zero (HbH and HbO₂H undissociated) to one (Hb⁻ and HbO₂⁻ completely dissociated). When $[Hb^-] = [HbH]$ or $[HbO_2^-] = [HbO_2H]$ the dissociation is 0.5 and pH = pK. It can be seen that pK_o is attained, when $[HbO_2^-] = [HbO_2H]$, at a pH value lower than pK_r, when $[Hb^-] = [HbH]$, indicating that HbO₂ is an acid stronger than Hb.

The difference in height of the two curves of Fig. 3.63 is plotted on the ordinate of Fig. 3.64, which represents the moles of H^+ set free at a given pH when Hb is oxygenated to HbO₂; this equals the moles of Na^+OH^- added to maintain the pH constant during oxygenation of Hb. This difference (*buffer capacity*) attains a maximum just at the physiological pH = 7.4 when 0.7 mol of H^+ are set free when

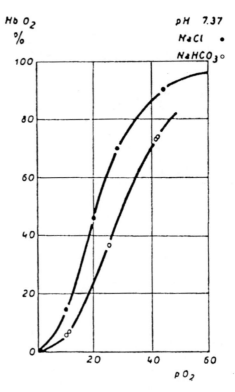

Fig. 3.62 Dissociation curves for oxygen (pO_2 in mmHg on the abscissa) of hemoglobin in solution of NaCl (filled points) and of sodium bicarbonate $NaHCO_3$ (open points). In both solution pH = 7.37 (from Margaria R and De Caro L *Principi di fisiologia umana*, Vallardi, Milano, 1977)

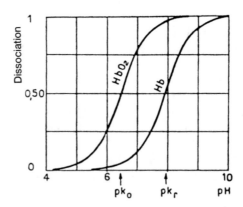

Fig. 3.63 Dissociation on the ordinate is: (i) the ratio $[HbO_2^-]/([HbO_2^-]+[HbO_2H])$ and (ii) $[Hb^-]/([Hb^-]+[HbH])$; these ratios are plotted as a function of pH on the abscissa: pH = pK_o + $\log([HbO_2^-]/[HbO_2H])$ on the left and pH = pK_r + $\log([Hb^-]/[HbH])$ on the right (from Margaria R and De Caro L *Principi di fisiologia umana*, Vallardi, Milano, 1977)

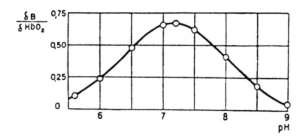

Fig. 3.64 Difference in height of the two curves in Fig. 3.63 indicating the *buffer capacity* i.e. the quantity of Na$^+$OH$^-$, which must be added to maintain constant the pH given on the abscissa when Hb oxygenates releasing H$^+$ (from Margaria R and De Caro L *Principi di fisiologia umana*, Vallardi, Milano, 1977)

one mole of HbH is oxygenated. This means that when the reduced blood reaches the alveoli and oxygenates: HbH $+$ 1O$_2$ $\leftrightarrows$ HbO$_2^-$ $+$ 0.7H$^+$. Since CO$_2$ $+$ H$_2$O $\leftrightarrows$ H$_2$CO$_3$ $\leftrightarrows$ H$^+$ $+$ HCO$_3^-$ the elimination of one molecule of CO$_2$ into alveolar air from blood will lose one hydrogen ion H$^+$ and this will be partially compensated by the release of 0.7 H$^+$ due to the oxygenation of one molecule of Hb. If the respiratory quotient QR $=$ 0.7 the compensation would be perfect.

In addition to CO$_2$ as such (Fig. 3.62) and *the pH (Bohr's effect)*, other factors affect the dissociation curve of blood for oxygen. Among these, the 2,3-diphosphoglycerate (2,3-DPG), which is present in human red blood cells competing with CO$_2$ for the same sites in the hemoglobin molecule. The 2,3-DPG binds with greater affinity to hemoglobin decreasing its affinity for oxygen, so it promotes the release of the remaining oxygen molecules bound to the hemoglobin, thus enhancing the ability of red blood cells to release oxygen near tissues that need it most. The 2,3-DPG increases in conditions of hypoxia as in high altitude when CO$_2$ decreases due to hyperventilation improving the release of oxygen to tissues. Another factor is the temperature: an increase in temperature moves the dissociation curve to the right (Fig. 3.65) i.e. decreases the affinity of oxygen for hemoglobin. This is particularly useful during muscular exercise because more oxygen is released by blood to the warm working muscles at the venous extremity of the capillaries. Therefore all factors which occur during muscular exercise: increased CO$_2$ production, increased temperature and eventually increase in [H$^+$] in anaerobic conditions (i.e. when the power output is greater than the maximal aerobic capacity) all displace the dissociation curve of blood for oxygen to the right facilitating the release of oxygen to the tissues.

The fetal hemoglobin has a dissociation curve for oxygen displaced to the left. This means that at a given pO$_2$ oxygen is delivered from the mother blood to the fetus. After birth there is a short lasting rebound before the curve attains normality.

As previously described (Fig. 3.46), the carbon monoxide CO has an affinity for hemoglobin 230 times greater than oxygen (Fig. 3.66). As shown in Fig. 3.51 in a human adult 100 ml of blood carry 20 ml of oxygen at saturation and its total blood volume, ~5 l, will carry 1 l of oxygen; it follows that 1 l of CO will occupy *all* the

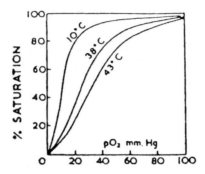

Fig. 3.65 Effect of temperature on the human blood oxygen dissociation curve determined at a pCO₂ of 40 mmHg (modified from Wright S *Applied Physiology*, Oxford University Press, London, 1955)

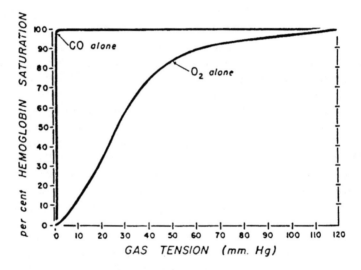

Fig. 3.66 Dissociation curves for HbO₂ and HbCO plotted on the same scale. Maximal saturation of hemoglobin with O₂ is not reached until the pO₂ is greater than 120 mmHg; with CO, however, maximal saturation is attained with a pCO of less than 1 mmHg (from Comroe JH, *The lung: clinical physiology and pulmonary function tests*, Chicago, Year Book Publishers, 1955)

sites (hemes) where oxygen can attach; in a 3 kg baby, the total volume of blood is ~200 ml and 200 ml of CO will be enough to block all his capabilities for oxygen transport.

Figure 3.67 shows the effect of partial amounts of CO on the oxygenation of the remaining CO-free hemoglobin. The dotted line is the dissociation curve of an anemic subject, having 40% of the total possible content of Hb, to be compared with that obtained by blocking 60% of the hemes of normal blood with CO leaving the other 40% free to bind O₂. It can be seen that whereas the total O₂ bound at saturation is about equal in the two conditions, the shape of the curves differs appreciably; in

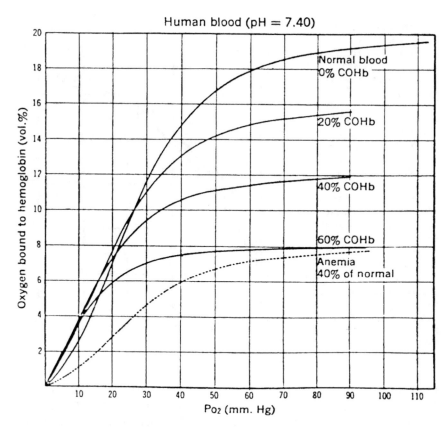

Fig. 3.67 Calculated O_2 dissociation curves of human blood containing various amounts of carboxyhemoglobin COHb, plotting absolute amount of O_2 content rather than percentage of available hemoglobin saturated with oxygen (as Fig. 3.61) (from Roughton FJW and Darling RC *American Journal of Physiology*, 141: 17–31, 1944)

particular the 60% COHb curve approaches that of myoglobin (Fig. 3.60) retaining O_2 just in the pO_2 range where oxygen is delivered to the tissues by normal blood (20–40 mmHg). The physiological conditions are therefore better in the anemic subject than in the subject partially poisoned with CO in spite of an equal total amount of sites where O_2 can bind.

In Fig. 3.68 is plotted the *area of dissociation of blood for oxygen*. The two thin lines delimiting the area (dashed) are two segments of the dissociation curves of blood for oxygen as those plotted in Fig. 3.61. As described above, the carbon dioxide CO_2 displaces the curve to the right (Fig. 3.61), it follows that there are two curves of dissociation of blood for oxygen: one for the arterial blood to the left and the other for the venous blood to the right. In the tissues, the blood coming from the lungs contains more O_2 and less CO_2 in the arterial extremity of the capillary, A in Fig. 3.68, and less O_2 and more CO_2 at its venous extremity, V in Fig. 3.68.

Fig. 3.68 Ordinates: Left, hemoglobin O_2 saturation, i.e. $[HbO_2/(Hb + HbO_2)] \times 100$; right, ml of O_2 in 100 ml of blood. Abscissa: partial O_2 pressure of oxygen in mmHg. The two thin lines delimiting the dashed area are sections of the Hb dissociation curves in the arterial (upper) and in the venous blood (lower). The two thick lines show the transient from the arterial blood to the venous blood (lower) along a capillary of the systemic circulation (A to V) and from the venous blood to the arterial blood along a capillary of the pulmonary circulation (V to A) (from Margaria R and De Caro L *Principi di fisiologia umana*, Vallardi, Milano, 1977)

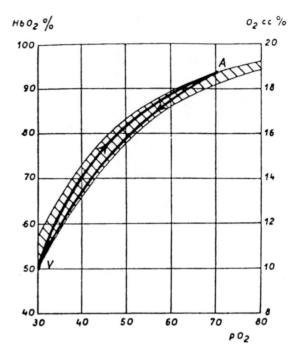

This implies a shift of the dissociation curve of O_2 from left to right (Fig. 3.61). This shift is physiologically useful because more oxygen, represented by the difference in height on the ordinate at the venous extremity V, is delivered to the tissues. The contrary is true at the level of the alveoli where more oxygen is acquired in the tract from the venous to the arterial end of the capillary. The anticlockwise sense of the two thick lines in Fig. 3.68 is explained by the fact that CO_2 diffuses first, from the upper to the lower thin lines, more quickly then O_2 due to its greater solubility in water (see Sect. 3.9).

3.14.2 Carbon Dioxide

Whereas the oxygen is transported only as physically dissolved (Henry's law) and bound to hemoglobin, the CO_2 is transported in five ways:

$$CO_2 + H_2O \leftrightarrows H_2CO_3 \leftrightarrows H^+ + HCO_3^- \leftrightarrows H^+ + CO_3^{--}$$
$$\text{and } R\text{-}NH_2 + CO_2 \leftrightarrows R\text{-}NHCOOH$$

What follows is aimed to compare the *relative* content of the five forms in which CO_2 is contained in blood: CO_2, H_2CO_3, HCO_3^-, CO_3^{--} and R-NHCOOH, where the fifth one is a carbamino compound bound to some protein radical R. Let's consider the relative importance of these five ways of transport.

The first is the CO_2 physically dissolved according to the Henry's law: *the solubility* of CO_2 *is 25 times greater than that of* O_2. At the physiological arterial blood pressure of $CO_2 = 40$ mmHg (Fig. 3.54a) we can calculate the ml of CO_2 physically dissolved as follows: 0.3 ml O_2: 100 mmHg $= x$ ml O_2: 40 mmHg; $x = (0.3 \times 40)/100 = 0.12$ ml O_2 and $0.12 \times 25 = 3$ ml *of* CO_2 *physically dissolved in* 100 ml *of blood at the* $pCO_2 = 40$ mmHg. *An equal amount of* 3 ml *of* CO_2 *is contained as carbamino compounds in* 100 ml *of blood at the* $pCO_2 = 40$ mmHg.

The *relative* importance of the other CO_2 compounds is determined as follows. First its *hydration*, which is catalyzed by the enzyme *carbonic anhydrase* present in red blood cells: $CO_2 + H_2O \leftrightarrows H_2CO_3$; the H_2CO_3 formation velocity is $v_1 = k_1[CO_2] \times [H_2O]$, that of dissociation is $v_2 = k_2[H_2CO_3]$. At equilibrium $v_1 = v_2$, i.e. $k_1[CO_2] \times [H_2O] = k_2[H_2CO_3]$ and $k_1/k_2 = K = [H_2CO_3]/([CO_2] \times [H_2O])$. The concentration of H_2O is so large in a water solution that $[H_2O]$ is practically constant (a mole of $H_2O = 18$ g; in 1 l 1000 g/18 g $= 55.5$ mol of H_2O/l). We can therefore write: $K \times [H_2O] = K_{hydric} = K^* = [H_2CO_3]/[CO_2] = 10^{-3}$: *in blood the* $[CO_2]$ *physically dissolved is* 1000 *times greater than* H_2CO_3. It follows that taking $H_2CO_3 = 1$, we have $CO_2 = 1000$ and $NHCOOH = 1000$. What is $[HCO_3^-]/[CO_2]$?

Let's consider the *first* dissociation of $H_2CO_3 \leftrightarrows H^+ + HCO_3^-$; $v_1 = k_1[H_2CO_3]$, $v_2 = k_2[H^+] \times [HCO_3^-]$. At equilibrium $v_1 = v_2$, i.e. $k_1[H_2CO_3] = k_2[H^+] \times [HCO_3^-]$ and $k_1/k_2 = K_1 = [H^+] \times [HCO_3^-]/[H_2CO_3] = 10^{-3}$. The last equation can be written as $1/[H^+] = (1/K_1) \times [HCO_3^-]/[H_2CO_3]$. Since, as shown above, $[H_2CO_3] \approx [CO_2] \times 10^{-3}$ we can write $1/[H^+] = (1/K_1) \times [HCO_3^-]/([CO_2] \times 10^{-3})$, i.e. $1/[H^+] = (1/10^{-6}) \times [HCO_3^-]/[CO_2]$ and in logarithmic form: $pH = pK_1^* + \log [HCO_3^-]/[CO_2] = 6.1 + \log [HCO_3^-]/[CO_2]$ (*Henderson-Hasselbalch equation*). The asterisk * indicates that is taken $[H_2CO_3] \approx [CO_2] \times 10^{-3}$. Often $pH = pK_1^* + \log [HCO_3^-]/[CO_2]$ is written as $pH = pK_1^* + \log [HCO_3^-]/\alpha \, pCO_2$ in agreement with Henry's law (Sect. 3.9). *At the physiological* pH $= 7.4$, the Henderson-Hasselbalch equation shows that $7.4 - 6.1 = 1.3 = \log [HCO_3^-]/[CO_2] = \log 20$. Since we took $H_2CO_3 = 1$, $CO_2 = 1000$ then $HCO_3^- = 20,000$.

{Just memorize that log on base 10 of a number is the exponent that must be given to 10 to obtain the number and that $\log 1 = 0$; $\log 2 = 0.3$; $\log 3 = 0.47$. You do not have to memorize anything else: in fact $\log 4 = \log (2 \times 2) = \log 2 + \log 2 = 0.6$; $\log 5 = \log 10/2 = \log 10 - \log 2 = 0.7$; $\log 6 = \log (2 \times 3) = \log 2 + \log 3 = 0.3 + 0.47 = 0.77$; $\log 7 = ? \sim 0.8$ ($=0.84$); $\log 8 = \log (2 \times 4) = \log 2 + \log 4 = 0.9$; $\log 9 = \log (3 \times 3) = 0.47 + 0.47 = 0.94$; $\log 10 = 1$; $\log 20 = \log 2 \times 10 = \log 2 + \log 10 = 0.3 + 1 = 1.3$; in particular remember the physiological condition: $pH = -\log [H^+] = -\log (4 \times 10^{-8}) = -(0.6 + (-8)) = -(-7.4) = 7.4$}

Let's now consider the second dissociation of the carbonic acid: $HCO_3^- \leftrightarrows H^+ + CO_3^{--}$ (CO_3^{--} is the *carbonate* ion). As usual: $v_1 = k_1[HCO_3^-]$ and $v_2 = k_2[H^+] \times [CO_3^{--}]$. At equilibrium $v_1 = v_2$, i.e. $k_1[HCO_3^-] = k_2[H^+] \times [CO_3^{--}]$

$$k_1/k_2 = K_2 = [H^+] \times [CO_3^{--}]/[HCO_3^-] = 1.8 \times 10^{-10}.$$

At $pH = 7.4([H^+] = 4 \times 10^{-8})$:

$$[CO_3^{--}]/[HCO_3^-] = K_2/[H^+] = (1.8 \times 10^{-10})/(4 \times 10^{-8})$$
$$= 0.45 \times 10^{-2} = 0.0045 = 1/222$$

For each $[CO_3^{--}]$ there are 222 $[HCO_3^-]$; since we took $[HCO_3^-] = 20{,}000$, it follows that $[CO_3^{--}] = 20{,}000/222 = 90$.

In conclusion, the *relative* content of CO_2 in blood are: $[H_2CO_3] = 1$; $[CO_2] = 1000$; $[NHCOOH] = 1000$; $[HCO_3^-] = 20{,}000$; $[CO_3^{--}] = 90$, i.e. at pH $= 7.4$ the ml of CO_2 on the ordinate of Fig. 3.69 are mainly in the form of bicarbonate HCO_3^-, the second form is $[CO_2]$ physically dissolved $= 5\%$ of $[HCO_3^-]$.

Let's now reconsider Fig. 3.63 showing the dissociations HbH $\leftrightarrows$ Hb$^-$ + H$^+$ and HbO$_2$H $\leftrightarrows$ HbO$_2^-$ + H$^+$ as a function of pH. The slope of these curves represents the *buffer capacity*: i.e. the dissociation/pH change, corresponding to the amount of base that must be added for unit change of pH (Fig. 3.64). *The buffer capacity is maximal when dissociation equals 0.5, i.e. when the concentration of dissociated molecules equals that of undissociated molecules, the ratio of the two concentrations equals unity and the log of their ratio equals zero. Returning to the Henderson-Hasselbalch equation, this occurs when* pH $= pK_1^* = 6.1$.

Figure 3.69 shows the dissociation curve of blood for CO_2.

The straight line at the bottom of Fig. 3.69 is the concentration of CO_2 physically dissolved according to the equation $[CO_2] = \alpha\ pCO_2$ (Henry's law) where α is 25 times greater for CO_2 than for O_2. Let's remember that $[CO_2] = 3$ ml/100 ml of blood at $pCO_2 = 40$ mmHg whereas $[O_2] = 0.3$ ml/100 ml of blood at $pO_2 = 100$ mmHg (Fig. 3.51). According to Fick's law (Sect. 3.9: $dm/dt = -D\,A\,\alpha\ \Delta pp/\Delta x$) the CO_2 diffuses 25 times faster than O_2.

The two curves *S.O.* and *S.R.* show that at a given partial pressure of CO_2 on the abscissa the content of CO_2 is greater in the venous blood than in the oxygenated blood. This is due to two factors: (i) the oxygenated hemoglobin is more acid than the reduced hemoglobin, i.e. HbH + O$_2$ $\leftrightarrows$ HbO$_2^-$ + H$^+$ (Bohr's effect, Fig. 3.63) and will decrease the hydration of CO_2 as shown by the equation $CO_2 + H_2O \leftrightarrows$ $H_2CO_3 \leftrightarrows H^+ + HCO_3^- \leftrightarrows H^+ + CO_3^{--}$; (ii) Fig. 3.62 shows that oxygen per se decreases the affinity of hemoglobin for CO_2 according to the equation HbCO$_2$ + O$_2$ $\leftrightarrows$ HbO$_2$ + CO$_2$ (we must keep in mind that O$_2$ binds to the Fe^{++} of the heme whereas CO_2 binds to the carbamino compound R-NH$_2$).

The two curves *V.P.R.* and *V.P.O.* indicate the concentration of CO_2 in the plasma *after* the reduced and the oxygenated blood had been equilibrated with the partial pressure of CO_2 given on the abscissa. Note that the curves of plasma contain more CO_2 than those of blood at a given pCO_2 on the abscissa. Why is this so? CO_2 hydrates faster in the red blood cells than in plasma because in the red cells is contained the *carbonic anhydrase*, and the concentration of carbamino compound R-NH$_2$ + CO$_2$$\leftrightarrows$R-NHCOOH is greater in the red cell due to the fact the concentration of Hb (33%) is greater than the concentration of proteins in plasma (6–8%). The question arises: if HCO_3^- is formed mainly in the red cell why its concentration is greater in plasma? The answer is the Donnan equilibrium (that we will consider later on) stating that the concentration of negative ion diffusible (HCO_3^-) is greater where

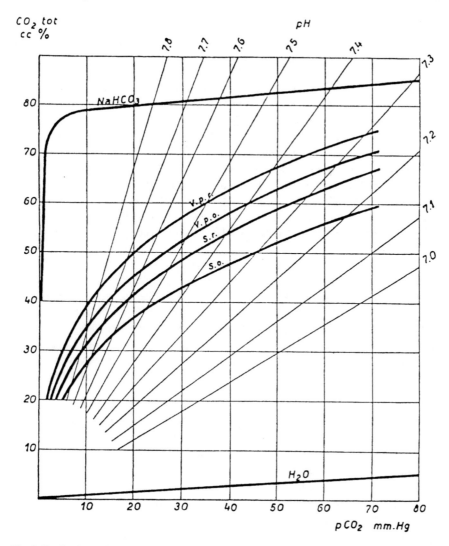

Fig. 3.69 On the ordinate are given the ml of CO_2 in 100 ml of arterial blood *S.O.*, venous blood *S.R.*, plasma of oxygenated blood *V.P.O.* and plasma of venous blood *V.P.R. separated from red cells after attainment of equilibrium with O_2*, as a function of the partial pressure of CO_2 in mmHg. The bottom straight line labeled H_2O indicates the CO_2 physically dissolved in distilled water according to Henry's law: in this case the fraction of CO_2 combined is only an unappreciable quantity of H_2CO_3. The thin oblique lines are iso-pH calculated from the Henderson-Hasselbalch equation by changing the values of $[HCO_3{}^-]$ and $[CO_2]$ without changing their ratio $[HCO_3{}^-]/[CO_2]$. The curve labeled $NaHCO_3$ is the dissociation curve for the CO_2 of a solution 0.0357 M of $NaHCO_3$ corresponding to a maximal capacity of combined CO_2 of 80 ml in 100 ml of solution (from Margaria R and De Caro L *Principi di fisiologia umana*, Vallardi, Milano, 1977)

the concentration of the negative ion non diffusible (Hb^-) is less. This explains why it is important to measure the concentration of CO_2 in the plasma *after the whole blood* is equilibrated with CO_2.

The upper bicarbonate curve $NaHCO_3$ shows that $[CO_2]$ increases sharply from 40 to 80 ml/100 ml of blood when the $p[CO_2]$ increases from 0 to 20 mmHg and then parallels the straight line: $[CO_2] = \alpha\, pCO_2$ (Henry's law) as that of CO_2 physically solved in water (bottom line in Fig. 3.69). When the pCO_2 is reduced (in our case from 20 mmHg to zero) the reduction of CO_2 in the following equilibrium: $Na^+ + HCO_3^- \leftrightarrows CO_2 + Na^+ + OH^-$, *sets free* $Na^+ + OH^-$ which reacts with the remaining bicarbonate according to the reaction: $Na^+ + HCO_3^- + Na^+ + OH^- \rightarrow H_2O + 2Na^+ + CO_3^{--}$, producing water plus Sodium Carbonate, Na_2CO_3, which is stable and releases CO_2 only under vacuum. At $pCO_2 = 20$ mmHg, each Na^+ is bound to one CO_2 (bicarbonate) whereas at $pCO_2 = 0$ mmHg each Na^+ is bound to one half CO_2 (carbonate); for this reason $[CO_2]$ is halved when pCO_2 is reduced from 20 mmHg to zero.

The question arises: why the bicarbonate curve crosses the ordinate at $pCO_2 = 0$ whereas the plasma and blood curves start from the origin? This is because in plasma and red cells the *proteins act as a buffer blocking the increase of* $[OH^-]$: $Na^+ + HCO_3^- + PrH \leftrightarrows CO_2 + Na^+ + OH^- + PrH \rightarrow CO_2 + Na^+ + Pr^- + H_2O$. Whereas in the upper $NaHCO_3$ curve the reduction of CO_2 sets free $Na^+ + OH^-$ which reacts with the remaining bicarbonate producing sodium carbonate Na_2CO_3, the weak acid proteins PrH in blood and plasma become relatively stronger when $[OH^-]$ increases and release their hydrogen ions H^+ which reacts with OH^- forming water and allowing $[CO_2]$ to decrease to zero when $pCO_2 = 0$.

Let's now consider the thin iso-pH straight lines in Fig. 3.69. We know that $pH = pK_1^* + \log[HCO_3^-]/[CO_2]$, i.e. $1/[H^+] = 1/K_1^* \times [HCO_3^-]/\alpha\, pCO_2$, which can be written as $[HCO_3^-] = K_1^*/[H^+] \times \alpha\, pCO_2$, which is the equation of a straight line starting from the origin having slope $K_1^*/[H^+]$, lower the greater the concentration of H^+, i.e. the lower the pH (Fig. 3.69). Note that in the above treatment we have considered the total CO_2 concentration on the ordinate equal to $[HCO_3^-]$ neglecting the other minor forms in which CO_2 is contained. *Alkaline reserve is defined as the total volume of* CO_2 *contained in* 100 ml *of blood when* $pCO_2 = 40$ mmHg (~50 ml *according to* Fig. 3.69). Why call *alkaline reserve* something deriving from the acid (H_2CO_3)? This is because CO_2 is contained in blood mainly as $NaHCO_3$; e.g. running above the maximal aerobic power produces lactic acid, HL : $Na^+ + HCO_3^- + H^+ + L^- \rightarrow Na^+ + L^- + H_2CO_3$, which thanks to the *carbonic anhydrase* dissociates into water H_2O and CO_2 which is exhaled. Note that the succession of reactions: $CO_2 + H_2O \leftrightarrows H_2CO_3 \leftrightarrows H^+ + HCO_3^- \leftrightarrows H^+ + CO_3^{--}$ and $R-NH_2 + CO_2 \leftrightarrows R-NHCOOH$ *does not* show the relative concentrations of the different CO_2 forms but rather the transit that *each molecule* must do to pass from one form to the other; imagine pedestrians crossing one street with a roundabout in the middle: *each* pedestrian must cross the street independent of the number of pedestrians staying on the roundabout.

Figure 3.70 shows the area of dissociation of blood for CO_2 to be compared with Fig. 3.68 showing the area of dissociation of O_2. One advantage of having an area

Fig. 3.70 Area of dissociation of CO_2. Ordinate: ml of CO_2 in 100 ml of blood. Abscissa: partial pressure of CO_2 in mmHg. The two thin lines delimiting the dashed area are sections of the dissociations curves of blood for CO_2 in the venous (upper) and in the arterial blood (lower). The two thick lines show the transient from the venous blood to the arterial blood along a capillary of the pulmonary circulation (V to A) and from the arterial blood to the venous blood along a capillary of the systemic circulation (A to V). The straight thin lines are iso-pH (from Margaria R and De Caro L *Principi di fisiologia umana*, Vallardi, Milano, 1977)

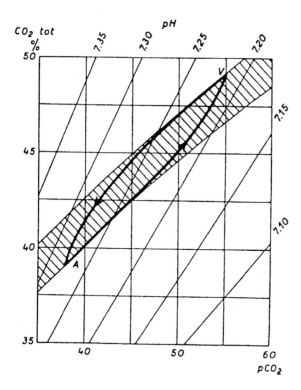

instead of a single curve is that more CO_2 can be assumed from the tissues into the blood from the arterial to the venous extremity of the capillary (A to V) given by the difference in height between the two curves at the extremity V. At the other extremity A the difference in height represents the additional CO_2 that can be delivered from the blood to the alveolar air. Furthermore the slope of the loop, i.e. that of a line joining A to V, crosses less iso-pH lines indicating a lower change in pH when CO_2 is absorbed/released by the blood along the capillary. The anti-clockwise direction of the two thick lines is explained by the fact that CO_2 diffuses faster then O_2, due to its greater solubility in water (see Sect. 3.9); it follows that CO_2 flows along its dissociation curve *before* the flow of oxygen causes a shift from one curve to the other.

Above we have seen the *relative* amount of the different forms of CO_2: $[H_2CO_3] = 1$; $[CO_2] = 1000$; $[NHCOOH] = 1000$; $[HCO_3^-] = 20,000$; $[CO_3^{--}] = 90$. Table 3.6 gives the *absolute* amount of the different forms of CO_2.

We have seen the large change in the *partial pressure* of O_2 relative to CO_2 in the transient from arterial blood and venous blood (Fig. 3.51). The carbamino compounds NHCOOH are already saturated with CO_2 in the arterial blood: i.e. the relationship between NHCOOH and pCO_2 in the arterial and venous pCO_2 physiological range is horizontal in both cases with a greater concentration of CO_2 in the venous blood than in the arterial blood. The two NHCOOH and pCO_2 curves of oxygenated and reduced

Table 3.6 The figures
indicate the milliliters of gas
in 100 ml of blood

	Art. Ven.
In solution	$3 + 0.5 = 3.5$ ml
Carbamino	$3 + 0.7 = 3.7$ ml
Bicarbonate	$42 + 2.8 = 44.8$ ml

Modified from Green JH *An Introduction to Human Physiology*,
Oxford University Press, London, 1968

blood are parallel. The difference in CO_2 acquired as NHCOOH in the arterial-venous blood transition, corresponds to 0.7 ml % of CO_2 picked up by the venous blood entirely due to a difference in the plateau attained by the two [NHCOOH]-pCO_2 curves, independent of the pCO_2; this CO_2 transfer is more rapid than that of CO_2 in solution and as bicarbonate. When describing the *relative* amount of the different forms of CO_2 we have seen that [HCO_3^-] is 20 times [CO_2] *at* pH $= 7.4$ (see above); however in the table above [CO_2] $= 3$ ml % and bicarbonate [HCO_3^-] $= 42$ ml % not 60 ml %. Blood is composed by the plasma and by the red cells where pH $= 7.1$. It follows that in the red cells: $7.1 = 6.1 + \log$ [HCO_3^-]/[CO_2]; $7.1 - 6.1 = 1 = \log 10$; i.e. [HCO_3^-]/[CO_2] $= 10$. Therefore it can be seen that in the red cell the ratio is 10 not 20, so if we had 100 ml of red cells only, then HCO_3^- will be $3 \times 10 = 30$ ml. If we had 100 ml of plasma only, where pH $= 7.4$, HCO_3^- will be $3 \times 20 = 60$ ml %. The total amount of HCO_3^- in 200 ml of blood (100 red cells $+$ 100 plasma) will be 90 ml, i.e. $90/2 = 45$ ml % a value similar to that given in the table above.

Figure 3.71 shows why in the plasma of the arterial blood the concentration of chlorine ions [Cl^-] is greater than in the plasma of the venous blood. When blood enters the capillary of the pulmonary alveolus, O_2 diffuses from air into plasma and in the red blood cell where it binds to the hemoglobin. As we have previously seen, oxygen: (i) decreases the affinity of hemoglobin for CO_2 and (ii) makes the Hb molecule more acid according to the equations: $HbCO_2 + O_2 \leftrightarrows HbO_2 + CO_2$ and $HbH + O_2 \leftrightarrows HbO_2^- + H^+$. Both these mechanisms result in a diffusion of CO_2 from the red cell into the plasma because: (i) the CO_2 previously bound to Hb gets free; and (ii) the H^+ combines with the bicarbonate ion producing carbonic acid H_2CO_3, which thanks to the *carbonic anhydrase*, contained in the red cell, dissociates into water and CO_2 which diffuse into the plasma. As shown in Fig. 3.71 the fast reduction of $H_2CO_3 \leftrightarrows H^+ + HCO_3^-$, due to the carbonic anhydrase, causes a fall of the negative ions HCO_3^- in the red cells compensated by an entry of HCO_3^- from the plasma. As shown in Fig. 3.71 this *entry* of negative ions HCO_3^- from the plasma into the red cell is compensated, by an *exit* of negative chlorine ions Cl^-, to maintain the same electrical potential in and out the red cell.

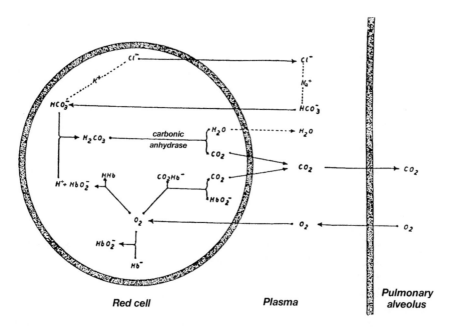

Fig. 3.71 Schema of the gaseous exchanges taking place in the lungs between alveolar air, plasma and red cells and of the corresponding chemical modification. Note that the loss of CO_2 implies an entry of HCO_3^- into the red cell from the plasma and an exit of negative chlorine ions Cl^- from the red cell to the plasma. The negative charges are compensated by positive sodium ions Na^+ in plasma and potassium ions K^+ in the red cell (from Margaria R and De Caro L *Principi di fisiologia umana*, Vallardi, Milano, 1977)

3.15 Acid-Base Equilibrium

Let's first remember the definition of $pH = \log 1/[H^+] = -\log [H^+]$; for example if $pH = 7.4$, the $[H^+]$ will be $7.4 = -\log (4 \times 10^{-8}) = -(\log 4 + (-8)) = -0.6 + 8 = 7.4$ (see how to memorize log of a number in Sect. 3.14.2). If the $[H^+]$ is halved: $pH = -(\log 2 \times 10^{-8}) = -(0.3 - 8) = 7.7$; if doubled: $pH = -(\log 8 \times 10^{-8}) = -(0.9 - 8) = 7.1$.

A *buffer solution* is a solution, which tends to maintain constant the pH in spite of an addition of OH^- or of H^+. We have already seen that the slope of Fig. 3.63 indicates the amount of base to be added for unit change of pH, i.e. the buffer capacity of a Hb solution attains a maximum when $pH = pK$. To obtain a buffer solution it is necessary: (i) to mix a weak acid with the same quantity of its salt with a strong base (e.g. CH_3COOH with $Na^+ CH_3COO^-$), or (ii) to mix a base with the same quantity of its salt with a strong acid (e.g. NH_3 with $NH_4^+ Cl^-$, or, when attached to an amino radical of a protein, $R\text{-}NH_2$ with $R\text{-}NH_3^+ Cl^-$).

In the first case, let's consider a weak acid (e.g. CH_3COOH); we can write the succession: $HA \leftrightarrows H^+ + A^-$; the arrow from left to right: $v_1 = k_1[HA]$, the other v_2

Table 3.7 From Green JH *An Introduction to Human Physiology*, Oxford University Press, London, 1968

Hemoglobin principal ionizing groups			
Group	No. of groups	Approximate pK	Principal form at pH 7.4
Terminal –COOH	4	4	– COO$^-$
Side-chain –COOH	56	4	– COO$^-$
Terminal –NH$_2$	4	7.7	– NH$_2$ and –NH$_3$$^+$
Side-chain –NH$_2$	44	11	– NH$_3$$^+$
Histidine $\equiv$N	38	7	$\equiv$N and $\equiv$NH$^+$

$= k_2[\text{H}^+] \times [\text{A}^-]$; at equilibrium $v_1 = v_2$, i.e. $k_1[\text{HA}] = k_2[\text{H}^+] \times [\text{A}^-]$; $k_1/k_2 = K = [\text{H}^+] \times [\text{A}^-]/[\text{HA}]$; $1/[\text{H}^+] = 1/K \times [\text{A}^-]/[\text{HA}]$; $\text{pH} = \text{pK} + \log[\text{A}^-]/[\text{HA}]$. If we had only [HA], i.e. CH_3COOH, the pH will be very low. To get $[\text{A}^-] = [\text{HA}]$, i.e. the maximum buffer capacity with $\text{pH} = \text{pK}$, we must add an equal amount of salt, e.g. Na^+ CH_3COO^- (salts are always dissociated), resulting in $[\text{CH}_3\text{COO}^-/[\text{CH}_3\text{COOH}] = 1$, $\log 1 = 0$ and $\text{pH} = \text{pK}$. In general: $\text{pH} = \text{pK} + \log\{[Salt:$ an acceptor of $\text{H}^+]/[Week$ *acid*: a donor of $\text{H}^+]\}$.

Let's now consider a base: $\text{NH}_4^+ \leftrightarrows \text{NH}_3 + \text{H}^+$ (or $\text{R-NH}_3^+ \leftrightarrows \text{R-NH}_2 + \text{H}^+$ when bound to an amino radical of a protein); the arrow from left to right: $v_1 = k_1[\text{NH}_4^+]$, the other $v_2 = k_2[\text{NH}_3] \times [\text{H}^+]$; at equilibrium $v_1 = v_2$, i.e. $k_1[\text{NH}_4^+] = k_2[\text{NH}_3] \times [\text{H}^+]$; $k_1/k_2 = K = [\text{NH}_3] \times [\text{H}^+]/[\text{NH}_4^+]$; $1/[\text{H}^+] = 1/K \times [\text{NH}_3]/[\text{NH}_4^+]$; $\text{pH} = \text{pK} + \log\{[\text{NH}_3]/[\text{NH}_4^+]\}$. Here: $\text{pH} = \text{pK} + \log\{[Base:$ acceptor of $\text{H}^+]/[Salt:$ donor of $\text{H}^+]\}$.

The different buffer systems of our organism are more efficient: (i) the more their pK is similar to the pH of its surrounding, and (ii) the greater their quantity.

Proteins are contained in great amount in plasma, and could be a buffer in the form of $\text{pH} = \text{pK} + \log[-\text{COO}^-]/[-\text{COOH}]$ and $\text{pH} = \text{pK} + \log[-\text{NH}_2]/[-\text{NH}_3^+]$. However their $\text{pK} = 5$–6, is smaller than the $\text{pH} = 7.4$ of their surrounding (Table 3.7), with the consequence that their numerator $[-\text{COO}^-]$ and $[-\text{NH}_2]$ will be larger than the denominator. On the contrary the hemoglobin, which is contained in great concentration in the red cells (33%) and in the blood (15%) contains an aminoacid in its molecule, which is a good buffer: the *histidine*.

It can be seen that in the hemoglobin molecule the *histidine* has a $\text{pK} = 7$ and an amount of 38 groups, making it a useful buffer system: $\text{pH} = 7 + \log[\equiv \text{N}]/[\equiv \text{NH}^+]$. For this reason the CO_2 entering in the capillary of the systemic circulation, first hydrates and quickly dissociates: $\text{CO}_2 + \text{H}_2\text{O} \leftrightarrows \text{H}_2\text{CO}_3 \leftrightarrows \text{H}^+ + \text{HCO}_3^-$ due to the combination of H^+ with the $\equiv$ N terminal of the histidine.

A second buffer system is that of phosphates particularly important in allowing the H^+ excretion by the kidney: $\text{pH} = 6.8 + \log[\text{HPO}_4^{--}]/[\text{H}_2\text{PO}_4^-]$, i.e. 7.4–$6.8 = 0.6 = \log 4$, therefore we have four $[\text{HPO}_4^{--}]$ for each $[\text{H}_2\text{PO}_4^-]$.

A third buffer system is: $\text{pH} = \text{pK}_1^* + \log[\text{HCO}_3^-]/[\text{CO}_2] = 6.1 + \log[\text{HCO}_3^-]/[\text{CO}_2]$ (Henderson-Hasselbalch equation). As described above (Sect. 3.14.2), during the hydration of CO_2, due to the *carbonic anhydrase*: $K =$

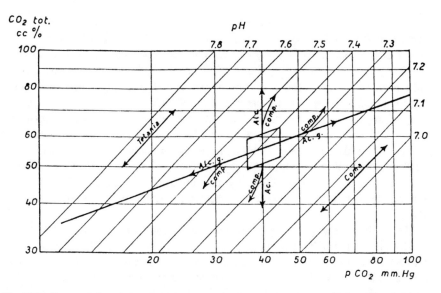

Fig. 3.72 Curve of dissociation for CO_2 of the plasma separated from red cells after attainment of equilibrium with O_2 (ordinate: ml of CO_2/100 ml of *v.p.o.* plasma in Fig. 3.69), as a function of the partial pressure of CO_2 in mmHg (abscissa). Both axis are given in a logarithmic scale instead of a linear scale as in Fig. 3.69 resulting in (i) a linear relationship of the CO_2 dissociation curve (at least within the limits of physiological interest) and (ii) parallel iso-pH lines with slope 45°. The area within the rectangle includes the physiological normal values. The straight arrows in vertical direction indicate acidosis (Ac. downward) and alkalosis (Alc. upward) due to a reduction, respectively and increase of the $[HCO_3^-]$ numerator term in the Henderson-Hasselbalch equation. The straight arrows along the CO_2 dissociation curve, indicate respiratory alkalosis (Alc. g. left) and respiratory acidosis (Ac. g. right) due a reduction, respectively an increase of the $[CO_2]$ denominator term in the Henderson-Hasselbalch equation. The curved arrows indicate the deviation caused by the compensation tending to attain normal values of pH; when compensation is insufficient and pH values attain ~7.75 *tetanus* ensues or when pH ~7.05 *coma* (from Margaria R and De Caro L *Principi di fisiologia umana,* Vallardi, Milano, 1977)

$[H_2CO_3]/([CO_2] \times [H_2O])$, the volume of H_2O in a water solution is so large that $[H_2O]$ is practically constant. The asterisk * indicates that $K \times [H_2O] = K_{hydric} = K^* = [H_2CO_3]/[CO_2] = 10^{-3}$ i.e. $[H_2CO_3] = [CO_2] \times 10^{-3}$.

Often $pH = pK_1^* + \log [HCO_3^-]/[CO_2]$ is written as $pH = pK_1^* + \log [HCO_3^-]/\alpha$ pCO_2 in agreement with Henry's law (Sect. 3.9). Note how important is the action of the carbonic anhydrase from the point of view of buffer capacity in changing pK_1 = 3 to $pK_1^* = 6.1$ (note that $K_1 = [H^+] \times [HCO_3^-]/[H_2CO_3] = 10^{-3}$ and $[H_2CO_3]$ $= [CO_2] \times 10^{-3}$). The Henderson-Hasselbalch equation shows that the pH can be adjusted with two mechanisms: the respiration by changing $[CO_2]$ and the kidney by changing the excretion of HCO_3^- (as we will see later on).

The straight line in Fig. 3.72 is empirically obtained by plotting the *v.p.o.* curve in Fig. 3.69 on a log–log scale. The parallel lines are iso-Ph. The area within the rectangle includes the physiological normal values.

The straight arrow to the left (Alc. g.) indicates what happens when in the Henderson-Hasselbalch equation (pH = 6.1 + log[HCO_3^-]/α pCO_2) the pCO_2 decreases because the subject hyperventilates, i.e. when the *alveolar* ventilation is greater than that required by the metabolism; this leads to respiratory alkalosis and the pH increases. If this condition is maintained, as at high altitude, the kidney compensates by increasing the excretion of HCO_3^- (downward *comp.* curve approaching the 7.4 pH line), but in doing so the *alkaline reserve* decreases (*alkaline reserve is defined as the total quantity of* CO_2 *contained in arterial blood when* $pCO_2 = 40$ mmHg: *normally* 48 ml *as indicated in* Table 3.6). The opposite case is the straight arrow to the right (Ac. g.), respiratory acidosis, due to an increase of pCO_2 ensuing respiratory insufficiency; the consequent decrease in pH is partially compensated (upward *comp.* curve approaching the 7.4 pH line) by kidney retaining HCO_3^-.

The vertical arrow pointing downward with a constant pCO_2 indicates, in the Henderson-Hasselbalch equation (pH = 6.1 + log [HCO_3^-]/[CO_2]) a decrease of [HCO_3^-], i.e. a metabolic acidosis due to the input of an acid in the circulation (e.g. lactic acid in an anaerobic exercise, or acetoacetic and beta-hydroxybutyric acid in diabetes). For example, the lactic acid: $H^+ + L^- + Na^+ + HCO_3^- \rightarrow Na^+ + L^- + H_2CO_3$ which dissociates in CO_2 and H_2O; the alkaline reserve decreases and the pH falls. Also in this case compensation is attained by increasing the ventilation (as indicated by the *comp.* arrow) in the attempt to maintain pH constant. In the hypothetic case that the two *comp.* curves meet (that departing from the *Alc. g.* line and that departing from the *Ac.* line, e.g. in a subject acclimated at high altitude and in a subject with diabetes) the physiological conditions in the graph of Fig. 3.72 would be the same in spite of their different anamnesis! The difference in the two cases is the *initial cause*: the decrease of the numerator [HCO_3^-] in case of the diabetes and the decrease of the denominator α pCO_2 in case of compensated alkalosis. The upward vertical arrow (*Alc.*) is a less frequent condition, e.g. the increase in [HCO_3^-] due to ingestion of bicarbonate; in this case the pH increases and the subject compensates with hypoventilation increasing the pCO_2.

3.16 Phonation

Unlike the vibrating lamina of an accordion or a hooter, in human phonation a given intensity of sound I_s (measured in Decibel, dB = 10 log I_s/I_{so} where $I_{so} = 10^{-16}$ W/cm^2) *can be attained with different flows* $\dot{V}$ *of the exhaled air* (Fig. 3.73). This is because in phonation two requirements must be simultaneously satisfied: sound production and breathing. A singer sacrifices breathing in favor of respiration when maintaining a note for a long time whereas in normal conversation the two requirements spontaneously meet.

Records in Figs. 3.73 and 3.75 were obtained as follows: a subject sat in a rigid body plethysmograph, which was airtight around the neck. Air was allowed to flow in and out the plethysmograph through a hole covered with Monel gauze whose

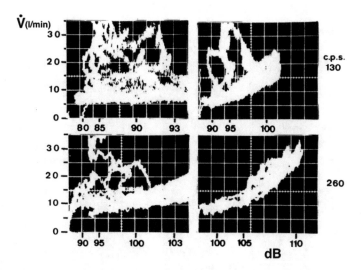

Fig. 3.73 Mean flow through the glottis $\dot{V}$ in liters per minute is recorded as a function of the sound level in dB on a *XY* storage oscilloscope while the subject held a tone of a constant pitch at different increasing and decreasing loudness. Two tones at the frequency indicated on the right (cycles per second) were tested. It can be seen that the same loudness dB could be maintained with very different airflows, *but that a minimum airflow* (bottom limit in each graph) is necessary to maintain a given intensity (modified from Cavagna GA and Margaria R *Ann. N.Y. Acad. Sci.* 155: 152, 1968)

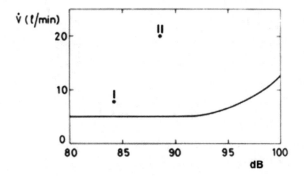

Fig. 3.74 The continuous line indicates the relationship between minimal airflow, $\dot{V}_{min}$, necessary to maintain a sound and the intensity of the sound emitted in dB. I and II indicate 130 Hz tones attained with a flow near (I) and much greater (II) than $\dot{V}_{min}$. One complete cycle of the vocal folds during sound emission in the two conditions is illustrated in Fig. 3.75 (from Cavagna GA and Camporesi EM in Wyke B (ed.) *Ventilatory and Phonatory Control System*, Oxford University Press, 1974)

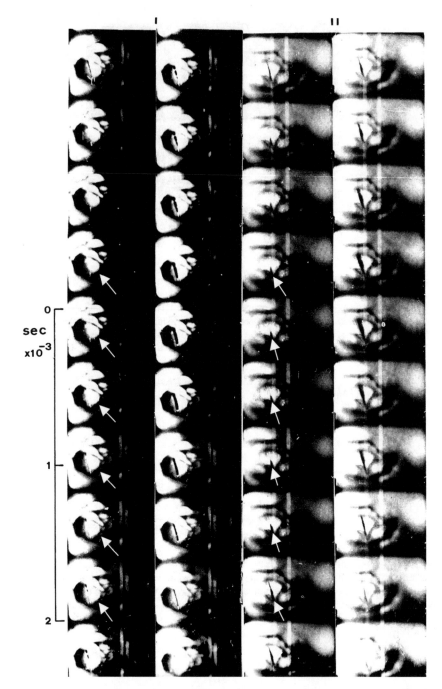

Fig. 3.75 Pictures of one cycle (two columns) of tones I (left) and II (right) in Fig. 3.74. The white arrows indicate that vocal folds close completely in I (lower flow) whereas the *posterior portion* of the larynx remains always open in II (greater flow) (modified from Cavagna GA and Camporesi EM in Wyke B (ed.) *Ventilatory and Phonatory Control System*, Oxford University Press, 1974)

resistance to airflow allowed measuring the flow $\dot{V}$ through the glottis. A microphone measured the intensity of sound production, dB. The signals corresponding to sound level dB and to the flow $\dot{V}$ were conveyed on the ordinate and on the abscissa respectively of an XY storage oscilloscope which could be seen by the subject during phonation, thus allowing voluntary regulation of the flow through the glottis during maintenance of a tone. A film of the vocal folds vibrating at 130 and 260 Hz taken at 2400 and 2300 frames/s was simultaneously recorded on a second oscilloscope with the corresponding $\dot{V}$ and dB levels (Fig. 3.75). Illumination of the vocal folds was supplied by a 7500 W lamp; in order to prevent 'sunburn' of the vocal folds, the red and infrared rays were in part reflected by a special glass. From Fig. 3.73 it can be seen that a given sound intensity (dB) can be attained with different airflows through the glottis, which can vary widely *but cannot be lower than a minimal value* $\dot{V}_{min}$ clearly shown by the bottom limit of the $\dot{V}$-dB records in Fig. 3.73. $\dot{V}_{min}$ is given as a function of dB by the continuous line in Fig. 3.74 for a 130 Hz sound maintained with two different airflows (I and II). The corresponding pictures of a vocal folds cycle filmed at 2400 frames/s are given in Fig. 3.75.

In conclusion: during phonation a compromise is attained between *sound production* and *breathing*. The efficiency of sound production is maximal when the vocal folds open and close completely during a vibrating cycle. Breathing requirement is afforded by a continuous opening of the cartilaginous posterior part of the glottis, oscillating at the same frequency, but without attaining a complete closure. An increased shunt of air through the posterior part of the glottis results in a 'breathy voice'.

Chapter 4
Kidney

Abstract General functions of kidney: maintaining constant the osmotic pressure of the organism, the relative concentration of different plasmatic components, the pH, the total isosmotic volume of liquid in the body, elimination of waste and producing hormones. Osmotic work. Anatomy of the nephron. Glomerular filtration by the proximal convoluted tubule. Forces determining the glomerular filtration and its auto regulation. Definition of clearance. Plasma clearance of inulin, para-aminohippuric acid and glucose. Mechanisms of the splay in the plot between mass of substance eliminated with urine in unit time and the plasmatic concentration of the substance. Clearance of the urea. Clearance changes of a substance versus its plasmatic concentration. Function of the proximal convoluted tubule, the Henle's loop, the distal convoluted tubule and the collecting tubule. Cycle of urea and osmotic diuresis. pH regulation by kidney: mechanism of reabsorption of filtered bicarbonate, and excretion of titration acid and ammonium sulphate. Water balance of our organism. Measurement and calculation of fluid compartments of the organism. Donnan equilibrium between compartments with diffusible and not diffusible ions through a separating membrane, resulting in a different ions concentration between adjacent compartments and an electrical potential difference across the membrane. Derivation of the Nernst equation.

4.1 Introduction

Kidney is one of the organs where exchanges occur between inside and outside of our body cooperating to maintain constant its internal composition. Whereas in muscle the functional unit is the half-sarcomere, in kidney the functional unit is the *nephron*. We have seen that the high-pressure side of blood circulation (arteries) is mostly connected with the low-pressure side (veins) by networks of capillaries set in parallel to each other so that blood flow in one district can be regulated without appreciable changes in the blood flow of other districts (Fig. 1.1). One exception is the capillary blood flow in the nephron where the *afferent arteriole* sprays in many capillaries in the renal glomerulus to join again into the *efferent arteriole* spraying in a net of capillaries around the tubule of the nephron, resulting into two nets of capillaries in

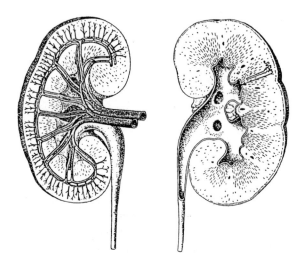

Fig. 4.1 Longitudinal section of a kidney. Left: arrangement of the blood vessels in the cortex and medulla. Right: topography of a nephron (from Margaria R and De Caro L *Principi di fisiologia umana*, Vallardi, Milano, 1977)

series. The nephron is made up by the *glomerulus* (*Bowman's capsule*), the *proximal convoluted tubule* followed by the *Henle's*, which descends into the medulla and then runs straight up again to the cortex where it continues as the *distal convoluted tubule* and the *collecting tubule* (Fig. 4.1).

The functions of the kidneys are:

(i) maintaining constant the *osmotic pressure* of the organism at its value 0.3 Osm, i.e. 0.3 Osm/l of solution. Remember that *osmolarity* is defined as moles of a substance/liter of *solution*, different from *osmolality* defined as moles/kg of solvent. One mole of a substance contains the Avogadro number of particles $= 6 \times 10^{23}$; for example 180 g of glucose contains the Avogadro number of molecules. Molarity depends upon the *number of molecules* in a solution, not on the identity of the molecules. For example, a 1 molar solution of a nonionizing substance such as glucose, i.e. 180 g in one liter of solution, is a 1 Osm solution; one molar solution of NaCl equals a 2 Osm solution

(ii) maintaining constant the *relative* concentration of the different molecules contributing to the total osmotic pressure

(iii) elimination of waste

(iv) contributing to maintain constant the pH of the organism

(v) maintaining constant the *total isosmotic volume of liquid* in the organism (as when we inject a solution with composition equal that of plasma). This last function is carried out by *receptors of volume* placed in the right atrium (in much less extent also in the left atrium), which release a substance called *atrial natriuretic factor, A.N.F.*, a peptide hormone secreted when the walls of the atrium, are stretched; the *A.N.F.* causes an increased diuresis by dilatation of the kidney arterioles resulting in a increased renal sodium excretion. Note that a distension of the right atrium walls may take place not only due to an increase of the body liquids, but also, as we have seen, in pathological conditions (as in the insufficiency of the right ventricle)

(vi) producing hormones: (1) the *renin*, which senses a low blood flow through the kidney and increases water retention; and (2) the *erythropoietin* which stimulates blood cells production in the bone marrow in response to hypoxia.

4.2 Osmotic Work

In the kidney we have the transformation of chemical energy into *osmotic work.* What is the osmotic work? *The Raoult's law states that the partial vapor pressure of each component of an ideal mixture of liquids is equal to the vapor pressure of the pure component multiplied by its mole fraction in the mixture.* A membrane that separates two solutions is defined *semipermeable* when it allows the passage of the solvent and not of the solute (for example water, but not glucose). In physiology is more convenient to consider *selectively* permeable membranes allowing the passage of the solvent and of *some* of the solute molecules.

In the example of Fig. 4.2 the force F over the surface area S of the piston is called *osmotic* pressure $\pi = F/S$ and is due to the different concentration of permeable molecules in the two compartments. According to Raoult's law the partial pressure of pure water (right) will be greater than that of water in a solution (left) because the concentration of water is greater on the right. Osmotic pressure opposes further transit of solvent into the solution. If we now push the piston with a pressure greater than π, the concentration of the solution increases as indicated by the curve in Fig. 4.3 while *positive osmotic work* is done: $W = \int \pi \, dV$ (the integral sign $\int$ represents the sum of rectangle areas given each by the product of an infinitesimal change in volume dV times the osmotic pressure π during this change, Fig. 4.3). *The curve in* Fig. 4.3 *is a hyperbola following Boyle's law*: $\pi V = nRT$, and the area below it has the dimensions of work (force time displacement: $\pi \times V = (F/l^2) \times l^3 = F \times l$). The amount of work done can be calculated as follows with the same procedure described in Sect. 3.11.2. The area of each infinitesimal small rectangle will represent an infinitesimal small amount of work done: $dW = \pi \times dV$; according to Boyle's law $\pi = nRT/V$, i.e. $dW = nRT \, dV/V$. As mentioned above, the only integral we have to remember in this course is $\int dx/x = \ln x + $ constant. It follows that $W = nRT \int dV/V = nRT \ln V + $ constant; *before work is done*: $0 = nRT \ln V_0 + $ constant, i.e. constant $= -nRT \ln V_0$ and $W = nRT \int dV/V = nRT \ln V - nRT \ln V_0 = nRT \ln V/V_0$. Since $V_0 > V$ the work W turns out to be negative when the force F in Fig. 4.3 makes positive work by increasing the concentration of the solution; this is because W is the work done by the osmotic pressure π which, left alone, would tend to expand the solution from V to V_0 making positive work. The *positive* work done by the kidney will then be

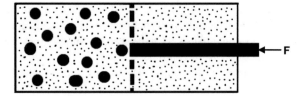

Fig. 4.2 The small molecules of water (indicated by the thin dots) can pass through the holes of the piston (membrane), whereas the large molecules (black circles) cannot. It follows that the piston will tend to move from left to right due to the fact that more molecules of water pass from right to left. To prevent this movement a force F must be exerted as indicated

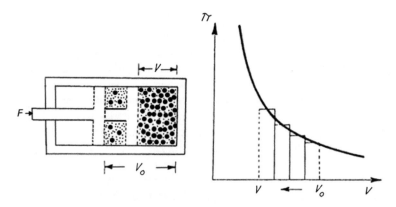

Fig. 4.3 Left: The force F pushes the piston from left (V_0) to right (V) and by doing so forces solvent molecules (small dots) to pass from right to left through the semipermeable membrane (interrupted line) and increases the concentration of the solute (large dots). Right: π–V diagram showing that when the volume is reduced from V_0 to V osmotic work is done given by the integral $\int \pi dV$, which is approximated by the sum of the areas of the rectangles drawn below the curve (see text)

$$W_{kidney} = nRT \ln V_{initial}/V_{final}$$
$$= nRT \ln \pi_{final}/\pi_{initial} (\text{since } \pi V = nRT = \text{constant} :$$
$$\pi_{initial} \times V_{initial} = \pi_{final} \times V_{final}, i.e. \ V_{initial}/V_{final} = \pi_{final}/\pi_{initial}).$$

When the kidney concentrates the urine (i.e. $\pi_{urine} > \pi_{plasma}$) performs positive work: this condition is called *hypersthenuria*. When $\pi_{urine} < \pi_{plasma}$ the condition is called *hyposthenuria*. When $\pi_{urine} = \pi_{plasma}$ the condition is called *isosthenuria*.

Often to measure the osmotic pressure of a solution it is used the *lowering of the freezing point*. The lowering of the freezing point of a 1 Osm solution (containing one mole of substance, i.e. 6×10^{23} particles per liter) is -1.86 °C; note that a mole of glucose per liter of water (180 g) is 1 Osm, whereas one mole of NaCl (58.44 g) per liter is 2 Osm because NaCl dissociates in water into two particles: Na^+ and Cl^-.

Table 4.1 Kidney's work done to excrete the indicated substances from plasma to 1 l of urine (from Margaria R and De Caro L *Principi di fisiologia umana*, Vallardi, Milano, 1977, after Borsook H and Winegarden HM *Proc Natl Acad Sci USA* 17:3, 1931)

Constituent	Plasma mol.	Urine mol.	Urine/Plasma	Work cal.
Na^+	0.135	0.152	1.125	11
K^+	0.00512	0.0384	7.5	48
Ca^{++}	0.00224	0.00375	1.67	1
Mg^{++}	0.00103	0.00347	3.37	1
Cl^-	0.104	0.166	1.60	48
$HPO_4^=$	0.000807	0.00326	4.05	3
$H_2PO_4^-$	0.000140	0.0125	89	35
SO_4^{--}	0.000312	0.0187	59.9	47
HCO_3^-	0.0266	0.00106	0.04	−2
Creatinine	0.000089	0.00664	74.5	18
Urea	0.00500	0.333	66.6	861
Urati	0.000238	0.00298	12.5	5
NH_4^+	0.000554	0.0222	40.0	50
Total				1126

Since *plasma* is 0.3 Osm, its freezing point will be $-1.86 \times 0.3 = -0.56$ °C. An example of the freezing point of urine could be -1.6 °C. Measuring the work in small calories ($R = 1.986$) in the equation $W_{kidney} = nRT \ln \pi_{final}/\pi_{initial}$, we can write (at 37 °C): $W_{kidney}(cal) = n \times 1.986 \times 310 \times 2.3 \log 1.6/0.56 = 1416 \times \log 1.6/0.56 \times n$. The number of moles n can be calculated, assuming a volume of urine of 1.5 l, 1:1.86 $= x{:}1.6$; i.e. $x = 1.6/1.86$ mol/l, i.e. $n = (1.6/1.86) \times 1.5$ and the work: $W_{kidney}(cal) = 1416 \times \log 1.6/0.56 \times (1.6/1.86) \times 1.5 = 833$ cal. (Note that $\ln_e x = 2.3 \log_{10} x$).

Table 4.1 shows that the work done by the kidney depends on both: (i) the *difference* in concentration of the constituent in the urine relatively to plasma, particularly large for example in creatinine and $H_2PO_4^-$, and (ii) the *amount* of substance excreted in the urine; both these factors are large in the *urea*, whose elimination represents most of the total kidney work.

4.3 The Nephron

To clean the inside of a bag you can overturn it, spread all its inside out and then put back in the objects you care about, leaving outside the garbage. This is the way the kidneys works: as shown in Figs. 4.4 and 4.5, 170 l of solution a day are allowed to pass at the level of the Bowman's capsule in the proximal tubule and then most of them, 168.5 l a day, are *selectively* absorbed along the nephron allowing only 1.5 l a day to be excreted as waste to the outside of our body.

Fig. 4.4 Components of the renal nephron (from Green JH *An Introduction to Human Physiology*, Oxford University Press, London, 1968)

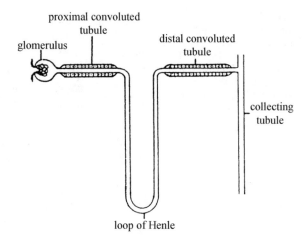

Fig. 4.5 Filtration and reabsorption of water by the kidney. This nephron represents all nephrons of both kidneys. In the illustrated normal conditions over 99% of the water filtered by the two kidneys is reabsorbed allowing 1 ml/min (~1.5 l a day) to reach the bladder for excretion (from Green JH *An Introduction to Human Physiology*, Oxford University Press, London, 1968)

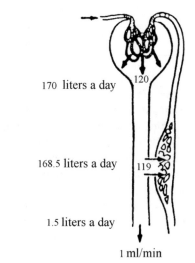

Let's now consider where these flows are coming from. The kidneys receive about 1300 ml of blood/min, i.e. ¼ of the cardiac output at rest (5 l/min); of these 1300 ml, red cells contribute 600 ml and plasma 700 ml. About 1/6 of the 700 ml of plasma are filtered (glomerular filtration ~120 ml/min, Fig. 4.5). Of these 120 ml/min, 7/8, i.e. 105 ml/min, are *always* reabsorbed by the proximal convoluted tubule (*obligatory reabsorption*), the other 15 ml continue along the nephron and are reabsorbed according to the physiological conditions (*optional reabsorption*). From these premises it follows that the *flow* in the proximal convoluted tubule is much faster than in the remaining portions of the nephron; for this reason the initial obligatory absorption is called *extensive* (more, but with less work to change the pH and the relative concentration of the consituents) whereas the subsequent is called *intensive* (less, but with more work to change the pH and the relative concentration of the consituents).

4.4   Glomerular Filtration

4.4.1   Characteristics of Glomerular Filtrate

Blood is made up by plasma and red cells, which obviously cannot pass through the capillary wall. Plasma is made up by water plus several salt ions and other small molecules (see Table 4.1), which easily pass through the capillaries wall, and relatively huge protein molecules, which do not filter (except *albumin*, the smallest of them, which can filter in small amount in some cases, particularly in blood stasis). The characteristics of the glomerular filtrate are determined using the apparatus shown in Fig. 4.6 by withdrawing a small amount of liquid for *analysis* and measuring its inside the *pressure* of the filtrate within the Bowman's capsule: this is done by measuring the height of the level bulb, which interrupts the flow of liquid within the pipette (e.g. in-out motion of a small air bubble). If the level of the bulb would equal that within the Bowman's capsule the pressure would be zero.

Table 4.2 shows some results of this procedure.

The *osmotic pressure* of the glomerular filtrate (0.3 Osm/l and −0.56 °C freezing point, corresponding to 5000 mmHg) practically equals that of plasma (5025 mmHg, where 25 mmHg is the oncotic pressure of the proteins), whereas its density is appreciably lower than that of plasma (1.007 vs. 1.025 g/ml). This is because the

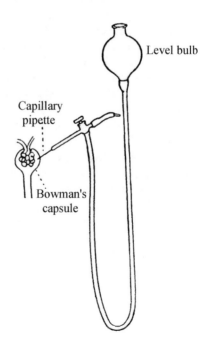

Fig. 4.6 Experimental set up to measure: (i) the *composition* of the glomerular filtrate in a small withdrawal of liquid (ii) the *pressure* of the filtrate within the Bowman's capsule: this is done by measuring the height of the level bulb (from Margaria R and De Caro L *Principi di fisiologia umana*, Vallardi, Milano, 1977, after Best CH and Taylor NB *The physiological basis of medical practice*, Bailliere Tindal & Cox, London, 1945)

Table 4.2 The first line indicates the Specific Gravity (a dimensionless unit defined as the ratio of the *density* of a substance to the density of water). The second and third lines refer to the osmotic pressure expressed as: (*i*) osmoles/liter (remember that 1 osmole contains the Avogadro's number, 6.02×10^{23}, of molecules), and (*ii*) lowering of the freezing point (from Green JH *An Introduction to Human Physiology*, Oxford University Press, London, 1968)

Sp. gr.	1.000	1.005	1.007	1.010	1.015	1.020	1.035
Osmoles/litre	0.0	0.2	0.3	0.4	0.6	0.8	1.4
Freezing point (°C)	0	−0.37	−0.56	−0.74	−1.1	−1.5	−2.6
				Glomerular filtrate			

osmotic pressure depends on the *number* of particles in solution whereas the density refer to the *mass* of the particles in solution: the small *number* of protein molecules in plasma increases little the osmotic pressure (25/5000 mmHg) but their large mass increases appreciably its density (1.025 vs. 1.007).

4.4.2   Forces Determining the Glomerular Filtration

In the capillaries of the systemic circulation (see Sect. 1.7.2), filtration takes place at their arterial extremity under the pressure gradient of 10 mmHg (35 mmHg hydraulic minus 25 mmHg oncotic pressure), and interstitial fluid is reabsorbed at the venous extremity under the same pressure of 10 mmHg (15 hydraulic−25 oncotic) (Fig. 1.39). In the capillaries of the pulmonary circulation, liquid is absorbed all along the alveolar capillary (10 hydraulic−25 oncotic). At the level of the capillaries within the Bowman's capsule, the filtration pressure $P_f = P_{cap} - P_{onc} - P_{Bowman}$, where P_{cap} is the hydraulic pressure within the capillary, P_{onc} is the oncotic pressure of the plasmatic proteins and P_{Bowman} is the 'back' hydraulic pressure of the liquid filtered within the Bowman capsule. In the equation above we know that $P_{onc} = 25$ mmHg, $P_{Bowman} \sim 15$ mmHg, measured as described above (Fig. 4.5), but we don't know P_{cap}. To do this we close the ureter and measure the increasing P_{Bowman} pressure until it stops at about 50 mmHg indicating that the filtration pressure P_f is zero. We can then write: $0 = P_{cap} - 25 - 50$, i.e. $P_{cap} = 75$ mmHg. It follows that in normal conditions $P_f = P_{cap} - P_{onc} - P_{Bowman} = 75 - 25 - 15 = 35$ mmHg. Note the high pressure in capillary within the Bowman capsule relative to that of the systemic circulation (25 mmHg), and pulmonary circulation (10 mmHg); this is due to: (i) the very short renal arteries connecting kidney to the aorta and (ii) the input section of the glomerulus artery slightly larger than the output section.

4.4.3 Auto Regulation

Figure 4.7 shows that renal blood flow and glomerular filtration are maintained constant in spite of large changes of arterial blood pressure. At blood pressures lower than about 50 mmHg the renal blood flow decreases sharply and this condition, if prolonged, may cause damage to the kidneys. The capacity of the kidney to work properly over a large range of arterial blood pressure is maintained even in the transplanted kidney indicating that it is largely independent of control by the nervous system.

The basic idea of the auto regulation *is to meet anatomically cause and effect*. This is shown in Fig. 4.8.

As shown in Fig. 4.5 the anatomical site where filtration takes place is positioned at the beginning of the nephron (glomerulus) whereas the anatomical site where most of the work has been done on the filtrate is positioned at the end of the nephron (distal convoluted tubule). Figure 4.8 shows that beginning and end of nephron meet at the *macula densa consisting in cells of the distal tubule* positioned at the angle between afferent and efferent arterioles. The macula densa position enables it to rapidly alter afferent arteriolar resistance in response to changes in the flow rate through the distal convoluted tubule. The macula densa uses the composition of the *distal tubular fluid* as an indicator of the glomerular filtration rate, GFR. As we will see, *large sodium chloride concentration is indicative of an elevated GFR, while low sodium*

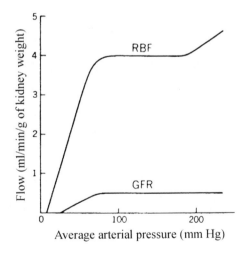

Fig. 4.7 Relationship between flow of a mammalian kidney and arterial pressure: it can be seen the auto regulation of renal blood flow (RBF) and glomerular filtration (GFR) in a perfusion range pressure of 80–180 mmHg (modified from Johnson RJ and Feehally J (*eds.*): *Comprehensive Clinical Nephrology*, Mosby, London, 2003)

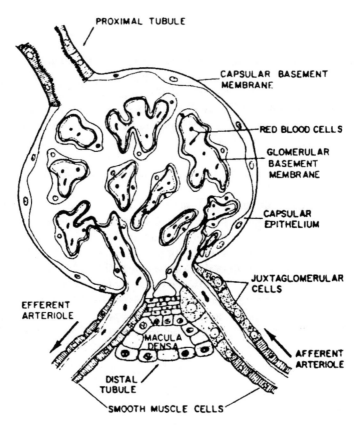

PROXIMAL TUBULE

CAPSULAR BASEMENT MEMBRANE

RED BLOOD CELLS

GLOMERULAR BASEMENT MEMBRANE

CAPSULAR EPITHELIUM

JUXTAGLOMERULAR CELLS

EFFERENT ARTERIOLE

MACULA DENSA

AFFERENT ARTERIOLE

DISTAL TUBULE

SMOOTH MUSCLE CELLS

Fig. 4.8 Diagram of the macula densa and of the juxtaglomerular complex (from Han AW and Leeson TS *Hystology*, Lippincott, Philadelphia, 1961)

chloride concentration indicates a depressed GFR. The cells of the macula densa are sensitive to the concentration of the sodium chloride in the distal convoluted tubule. In particular: the mechanisms that participate in kidney auto regulation are: (i) the *macula densa cells* sensing the Na^+Cl^- *content in the distal tube; in case this is too low because the flow is too small, these cells by means of excretion of prostaglandins* stimulate locally *the juxtaglomerular cells* to set free *renin* whose primary function in the kidneys is to increase blood pressure, leading to restoration of perfusion pressure; (ii) the *juxtaglomerular cells*, distinguished by their granulated cytoplasm (Fig. 4.8), which are positioned in the wall of the afferent arteriole and release *renin* when relax because blood pressure falls.

4.4.4 Clearance

In words: *clearance* is the volume of *plasma* set free from a given substance in its passage through the kidney. This is shown in Fig. 4.9.

Obviously the volume of plasma cleared by a substance does not exit from the kidney separated from the polluted ones (as schematically shown by the white square in Fig. 4.9), but mixed within the whole volume of plasma output.

The same argument can be quantitatively expressed as follows: the quantity of substance excreted in the urine in unit time $\dot{Q}_u$ equals that which has left the plasma, $\dot{Q}_p$, i.e. $\dot{Q}_u = \dot{Q}_p$. But $\dot{Q} = C \times \dot{V}$ where C equals the concentration of a substance in a volume of liquid and $\dot{V}$ is the flow of liquid where the substance is dissolved. It follows that $U \times \dot{V}_u = P \times \dot{V}_p$ where U is the concentration of the substance in the volume of urine excreted in unit time $\dot{V}_u$ and P is the concentration of the substance in the volume of plasma $\dot{V}_p$ depurated by the substance in the same time. *It follows the definition of clearance:* $\dot{V}_p = U \times \dot{V}_u / P$. Substances differ according to their clearance. *High-threshold* substances (such as chlorides or glucose) are entirely reabsorbed: no clearance. *Low-threshold* substances (such as phosphates or urea) are reabsorbed in limited quantities: low clearance. *No-threshold* substances (such as para-aminohippuric acid, PAH) are totally excreted: high clearance.

The polysaccharide *inulin* is not naturally found in human plasma, it is nontoxic and can be injected into the bloodstream; this substance has the characteristics to filtrate freely through the glomerulus and not to be absorbed or excreted by the renal tubules (Fig. 4.10 left diagram). All the mass of inulin filtered through the glomerulus is found in the urine and equals $\dot{V}_u \times U$ where $\dot{V}_u$ is the volume of urine excreted in unit time and *U is the concentration of inulin in the urine*. This mass of inulin in the urine equals the mass of inulin that has left the plasma in the same time: $\dot{V}_p \times P$ where $\dot{V}_p$ is the volume of plasma set free (cleared) by inulin, and *P is the concentration of inulin in the plasma*. In symbols: $\dot{V}_u \times U = \dot{V}_p \times P$, in case $\dot{V}_p$ is constant this is the equation of a straight line starting from the origin having $\dot{V}_u \times U$ on the ordinate, P on the abscissa and slope $\dot{V}_p$, i.e. *the flow of plasma cleared by the substance.* In case of the inulin, it is found experimentally a straight line starting from the origin with slope $\dot{V}_p = 120$ ml/min (left graph in Fig. 4.10): the *clearance of the inulin equals the glomerular filtration rate* (Fig. 4.5).

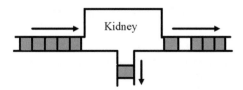

Fig. 4.9 Schematic representation of plasma clearance: of five ideal volumes of plasma polluted with a substance (left), one volume of plasma is cleared from the substance by the kidney (white square on the right) and its content is eliminated with the urine

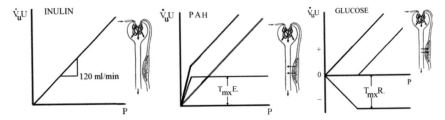

Fig. 4.10 The mass of substance eliminated with urine in unit time $\dot{V}_u \times U$ is plotted on the ordinate as a function of the concentration P (abscissa) of the substance in the volume of plasma $\dot{V}_p$ set free by the substance. Since $\dot{V}_u \times U = \dot{V}_p \times P$ the relationship is a straight line starting from the origin with slope equal to $\dot{V}_p$. The graph on the left refers to *inulin*, a substance which filters freely through the glomerulus and is subsequently eliminated into urine without exchanges with the blood vessels surrounding the tubules; in this case the *slope* $\dot{V}_p = 120$ ml/min *equals the glomerulus filtration rate* (Fig. 4.5). The graph on the center refers to the para-aminohippuric acid, *(PAH)* which, as inulin, filters freely through the glomerulus, but in addition is excreted from blood vessels into the tubules (as indicated by the two arrows) increasing with P to a maximum (*maximum excretory load:* T_{mxE}.) after which active excretion remains constant; the total mass of *PAH* found in the urine up to this point (upper line) is the sum of that filtered passively plus that excreted actively before T_{mxE}. is attained, the slope of the resulting line is the *total flow of plasma through the kidneys set free from PAH*; if plasmatic concentration increases further *PAH* is eliminated into the urine at the same rate of inulin. The graph on the right refers to *glucose*, which is actively reabsorbed by the tubules (as indicated by the two arrows) up to a limit (*maximum reabsorption load:* T_{mxR}.) after which is eliminated into the urine at the same rate of inulin (modified from Keele CA and Neil E *Samson Wright's applied physiology*, Oxford University Press, London, 1971)

Some substances, as for example the para-aminohippuric acid, PAH, not only filter freely as inulin, but are also *actively excreted* by the tubular cells (middle graph of Fig. 4.10); in other words, the ordinate of Fig. 4.10 refer to both a *passive* and an *active* mechanism. The passive mechanism equals that of inulin resulting in a linear relation starting from the origin and increasing with no limits with increasing concentration of the substance in the plasma. On the contrary the excretion of actively transported substances, such as the PAH, attains a limit above which the amount of substance excreted in the tubules of the nephron remains constant because all the 'transporter' molecules are saturated. Before this limit is attained some of the filtrated plasma continues its flow through the renal tubule with no more substance available to be excreted actively, i.e. the *clearance of PAH up to this concentration represents all the plasma flow through the kidney*: ~650 ml/min corresponding to a blood flow of about 1200 ml/min.

The right hand graph in Fig. 4.10 refers to a substance with a high threshold such as the glucose, i.e. a substance which normally is retained by the organism. Also in this case two mechanisms take place: a passive one, such as that of inulin, indicated by the straight line starting from the origin towards positive values of mass of glucose filtered into the nephron tubule $\dot{V}_p \times U$, and an active one which is *reversed* indicating an active transport from the tubule back into the blood vessels (negative values of $\dot{V}_u \times U$). It can be seen that up to a given value of glucose concentration in the plasma (P on the abscissa) the lines in the positive and negative range of the ordinate are

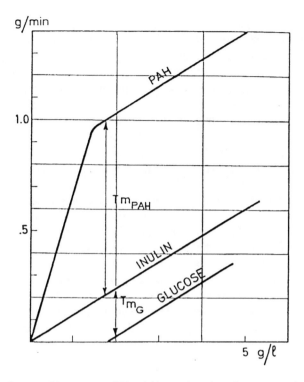

Fig. 4.11 The three resulting curves of Fig. 4.10 are plotted on the same graph with quantitative values on the ordinate and the abscissa. Note that glucose appears in the urine at a plasmatic concentration of ~2 g/l: the vertical arrows *Tmg* indicates the maximum reabsorption load of glucose: ~0.2 g/min. The vertical arrow *TmPAH* indicates the maximum excretory load of the para-aminohippuric acid ~780 g/min, the slope of the PAH line starting from the origin up *TmPAH* is ~660 ml/min. The slope of inulin is 120 ml/min (modified from Margaria R and De Caro L *Principi di Fisiologia Umana*, Francesco Vallardi, Milano, 1977)

specular, i.e. *all* the glucose filtered is actively reabsorbed: $\dot{V}_u \times U = 0$, the *clearance is zero*. Above this *maximum reabsorption load* T_{mxR} the active mechanism saturates and all the glucose filtered is lost in the urine (Fig. 4.11).

4.4.5 Splay

The lines drawn in Fig. 4.10 start and end with sharp angles. In reality they start and end progressively, i.e. with a curvature more or less marked called *splay*. There are two reasons for this: *anatomical* and *physico-chemical*. The anatomical mechanism is easily understood considering that not all nephrons are exactly equal: some have a large glomerulus and a short tubule and other a smaller glomerulus and longer tubule. Consider, for example, the PAH: in the first type of glomerulus the saturation of the molecular transporters will take place at a lower concentration of PAH in the plasma (abscissa). The contrary will be true for a nephron with a smaller glomerulus and longer tubule. All the anatomically different glomeruli will attain their *maximum excretory load* at different P concentrations with the results that the sum of all of them will approach T_{mxE} not simultaneously but with a gradual trend (splay).

To understand the second mechanism we have to consider that the absorption (e.g. glucose) and excretion (e.g. PAH) take place by binding molecules flowing into the tubule with a *carrier molecule* on the tubule wall. Let's consider, e.g. the reabsorption of glucose G binding to a carrier molecule T. As usual we will have: $T + G \rightleftarrows TG$; $v_1 = k_1[T][G]$ and $v_2 = k_2[TG]$; at equilibrium $v_1 = v_2$, i.e. $k_1[T][G] = k_2[TG]$ and $k_1/k_2 = K = [TG]/[T][G]$, where K, is the equilibrium constant. In order to have an instantaneous production of $[TG]$ with no presence of the reactants $[T][G]$ we should have $K = \infty$, which is nonsense. This means that flow of some molecules of G in the tubule will continue even if there are some free carriers T at disposal in the tubule wall, i.e. that the *maximum reabsorption load* T_{mxR} is not fully completed at a given value of glucose plasmatic concentration P. The cycle must be repeated in subsequent filtrations with higher glucose concentration P before T_{mxR} is reached. It must be pointed out that the *splay* is not equal for all the substances; the curvature can be prolonged for a large range of plasmatic concentration P; the tracings in Fig. 4.10 are just examples of the three possible ways substances can be eliminated into the urine.

4.4.6   Urea Clearance

Since *urea* is physiologically present in plasma, its clearance is clinically used to assess kidney function without the need to inject a test substance (e.g. inulin).

A low clearance of the urea could be due to: (i) a kidney disease and/or (ii) to a low urine flow as shown in Fig. 4.12. In order to distinguish between these two possibilities the *normal relationship* between urea clearance $\dot{V}_p = \dot{V}_u \times U/P$ and urine flow $\dot{V}_u$ has been approximated by the equation $\dot{V}_p = k\sqrt{\dot{V}_u}$ in the range $\dot{V}_u = 0$–2 ml/min and a *standard clearance of the urea* $\dot{V}_p^*$ has been defined for a urine flow $\dot{V}_u = 1$ ml/min corresponding to a clearance $\dot{V}_p = 54$ ml/min as a normal value (Fig. 4.12): $\dot{V}_p^* = k\sqrt{\dot{V}_u} = k\sqrt{1} = k$, i.e. $\dot{V}_p = \dot{V}_p^* \times \sqrt{\dot{V}_u}$, or $\dot{V}_p^* = \dot{V}_p/\sqrt{\dot{V}_u} = \dot{V}_u U/(P \times \sqrt{\dot{V}_u})$. It follows that $\dot{V}_p^* = \sqrt{\dot{V}_u}\sqrt{\dot{V}_u}U/(P \times \sqrt{\dot{V}_u}) = \sqrt{\dot{V}_u}U/P$ which *is the clearance standard of the urea.* This procedure allows to distinguish between the two possibilities mentioned above in case of a low urea clearance: a value lower than $\sqrt{\dot{V}_u}U/P = 54$ ml/min will not be due to a low urine flow, but to a kidney disease.

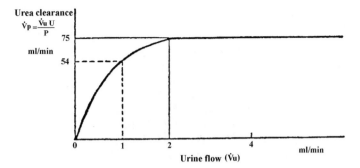

Fig. 4.12 Normal relationship between clearance of urea and urine flow. The interrupted line indicates the usual condition, $\dot{V}_p = 54$ ml/min corresponding to a urine flow of 1 ml/min (Fig. 4.5). A maximum of clearance, 75 ml/min of plasma set free from urea, is attained at a urine flow of 2 ml/min. Up to this point the relationship between $\dot{V}_p$ and $\dot{V}_u$ can be analytically described as indicated in the text (from Green JH *An Introduction to Human Physiology*, Oxford University Press, London, 1968)

4.4.7 Clearance Changes of a Substance Versus Its Plasmatic Concentration

Up to now (Figs. 4.10 and 4.11) we have considered the mass of substance found in unit time in the urine $\dot{V}_u U$ *as a function of its concentration in the plasma* P: this allowed us to define the *clearance* of a substance: $\dot{V}_u U / P$. The question now arises: how will the *clearance* $\dot{V}_u U / P$ change as a function of P? For example, Fig. 4.10 shows that the clearance of inulin will not change with P because $\dot{V}_u U = k P$, i.e. $\dot{V}_u U / P = k$, but this is not the case for other substances (e.g. PAH and glucose, Fig. 4.10). We will define: (i) $\dot{V}_u U$ the *excreted load*, i.e. the grams of substance found in urine in unit time, (ii) *filtered load* the grams of substance freely filtered through the glomerulus in unit time and (iii) T the *tubular load: the grams of substance excreted or absorbed, actively, passively or both by the nephrons tubules in unit time*. The *excreted load = the filtered load ± the tubular load*, i.e. $\dot{V}_u U = C_i \times P \pm T$ where C_i is the clearance of *inulin*, i.e. the flow of plasma through the glomerulus (left graph in Fig. 4.10); dividing both terms of the equation by P we obtain $\dot{V}_u U / P = C_x = C_i \pm T/P$ where C_x is the clearance of any substance. In case of *inulin* $T = 0$, i.e. $C_x = C_i$ independent of P.

In case of *PAH*: $C_{pah} = C_i + T/P$. If P is lower than the critical value $P_{c,pah}$ where T_{mxE} is attained (middle graph in Fig. 4.10), $T = kP$ i.e. $C_{pah} = C_i + kP/P = C_i + k = 720$ ml/min. Above the critical value $P_{c,pah}$, the maximum excretory load T_{mxE} is attained and maintained constant with increasing P, i.e. $C_{pah} = C_i + T_{mxE}P$: with increasing P the clearance of PAH tends to that of inulin (Fig. 4.13).

In case of *glucose* (right graph in Fig. 4.10) *all* (could be less, but not more) the filtered glucose is actively and completely reabsorbed up to its critical plasmatic concentration $P_{c,gluc}$ when T_{mxR} is attained, i.e.: $C_{gluc} = C_i - C_i P/P = 0$. Beyond

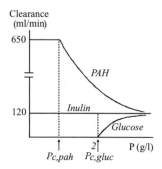

Fig. 4.13 Schematic representation of the clearance of the para-aminohippuric acid *PAH*, of *Glucose* and of *Inulin* as a function of their plasmatic concentration *P*. The arrows indicate the plasmatic concentrations $P_{c,pah}$ and $P_{c,gluc}$ where the maximal excretory load of *PAH* and the maximal absorption load of *Glucose* are attained. It can be seen that with increasing *P* all clearances tend to that of *Inulin* as described in the text

$P_{c,gluc}$, glucose is excreted in the urine: $C_{gluc} = C_i - T_{mxR}/P$, i.e. with increasing P the clearance of glucose tends to that of inulin (Fig. 4.13).

At the threshold value $C_{gluc} = 0 = C_i - T_{mxR}/P_{c,gluc}$; i.e. $T_{mxR}/P_{c,gluc} = C_i$ and $P_{c,gluc} = T_{mxR}/C_i$. This is a very important relationship: the plasmatic concentration of glucose $P_{c,gluc}$ beyond which glucose appears in the urine increases with decreasing kidney's glomerular filtration C_i. Clinically: a diabetic subject with a very high glucose concentration in the blood may not have glucose in the urine if his kidneys are affected by a pathology (e.g. a nephritis) resulting in a decrease of C_i. The normal values are $P_{c,gluc} = T_{mxR}/C_i \approx 375$ mg min^{-1}/120 ml min$^{-1} \approx 3$ mg/ml. Actually in normal conditions glucose appears in the urine at ~2 mg/ml (Fig. 4.11) because T_{mxR} is attained with a *splay* (Sect. 4.4.5) (Fig. 4.14).

4.5   Proximal Convoluted Tubule

The example previously described (Fig. 4.5) that in order to clean the inside of a bag you can overturn it, spread all its inside out and then put back in the objects you care of, leaving outside the garbage, applies well to the proximal convoluted tubule.

The flow of filtrate decreases drastically in the proximal tubule, 7/8 of it is *obligatory reabsorbed*, i.e. *of the 120 ml/min filtrate by the glomerulus only 15 ml/min flow at the end of it.* As mentioned in the Introduction one function of kidney is to maintain constant the volume of liquids in our organism (preventing *expanded extracellular fluid*) under the influence of the *atrial natriuretic factor ANF* synthesized and secreted by cardiac muscle cells in the walls of the atria in the heart. This is done by increasing sodium chloride excretion (as shown in Table 4.1 sodium chloride Na^+Cl^- is the principal molar salt constituent of urine) while maintaining constant the *relative* concentration of normal, physiological, components within the plasma

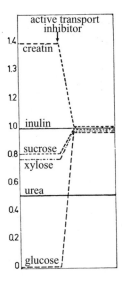

Fig. 4.14 Clearance relative to inulin of several substances in human body: normal and after injection of an active transport inhibitor (arrow). Note that substances having a clearance greater than that of inulin are reabsorbed, vice versa those with a lower clearance are excreted. Urea is unaffected because its clearance is passive (modified from Smith HW *The physiology of the kidney*, Oxford University Press, New York, 1937)

and excreting the abnormal components. All this takes place in the proximal tubule by selective absorption of all the substances filtered by the glomerulus.

At the end of the proximal convoluted tubule the osmotic pressure of the primitive urine about equals that of plasma (0.3 Osm), the pH $= pK + \log ([HCO_3{}^-]/\alpha pCO_2)$ does not appreciably differ from that of plasma: since $[CO_2]$ is the same (due to free gas diffusion) this means that practically all (~90%) of $HCO_3{}^-$ is absorbed by the proximal convoluted tubule.

The selective absorption of all the substances filtered by the glomerulus takes place by means of two kind of transfer: *active* and '*passive*' (this is between brackets because it is the consequence of the active mechanism). The *active* transport in turn takes place by means of two mechanisms: the Na^+ *pump* and *carriers* (e.g. glucose absorption and PAH excretion). *If 7/8 of glomerular filtrate are reabsorbed by the proximal tubule, this is due to active absorption of Na^+, which, as mentioned above* (Table 4.1) *represents the greater molar solute in plasma. Water, Cl^-, urea, $HCO_3{}^-$* and other constituents *will follow 'passively' due to the active reabsorption of Na^+* (Fig. 4.14). The *carriers* are molecules, which bind to a substance and then diffuse towards the interstitial liquid (absorption) or the tubular lumen (excretion); their existence has been shown by the phenomenon of *competition*. For the *absorption* transporters competition is demonstrated by the fact that if we inject fructose together with glucose we observe that the plasmatic threshold of glucose ($P_{c,gluc}$ in Fig. 4.13) increases; this is because some of the carriers are occupied to transport fructose

PROXIMAL TUBULE

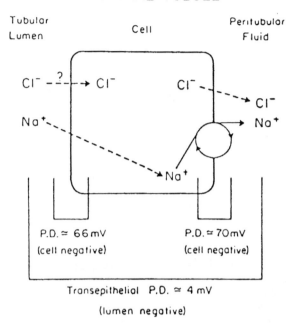

Tubular   Peritubular
Lumen   Cell   Fluid

P.D. ≃ 66 mV   P.D. ≃ 70 mV
(cell negative)   (cell negative)

Transepithelial   P.D. ≃ 4 mV

(lumen negative)

Fig. 4.15 The glomerulus filtrate flows at high speed (progressively decreasing) on the left side of the proximal tubule cell and most of it is reabsorbed by the cell and released to the peritubular fluid (right side of the cell). The electrical potential is zero in the peritubular fluid, −70 mV within the cell and −4 mV within the lumen. The interrupted lines indicate passive transport, the continuous circled line indicate the active pump transferring positive sodium ions Na^+ into the peritubular fluid against an electrical potential. The entry of Cl^- into the cell is signed with a question mark possibly suggesting some type of carrier allowing its entry against a potential difference (from Valtin H *Renal function*, Little Brown and Company, 1973)

from tubular lumen into blood, less carriers are free to transport glucose from tubule to blood with the result that glucose will appear in urine at a greater plasmatic concentration due to a lesser active absorption from tubule to interstitial fluid. The same is true for two kinds of *excretion* transporters: those excreting PAH, penicillin and other substances, and those excreting histamine, thiamine and other substances.

The flow of the glomerular filtrate rapidly decreases along the proximal tubule thanks to an *active* pump transferring Na^+ from inside the tubular cell into the peritubular space (Fig. 4.15). This process is initiated by the passive entry of Na^+ from lumen into the cell due to an electrochemical potential difference: the negative potential (−70 mV) and the lower concentration of Na^+ inside the cell (the entry of Na^+ from lumen into the cell makes the lumen potential slightly negative −4 mV, i.e. the actual potential difference pushing Na^+ into the cell is −66 mV). The Na^+ is then *'pumped' actively* from inside the tubular cell into the blood by an electrogenic pump against a 70 mV potential difference. Whereas the active transport by *carriers* attains a limit (Fig. 4.13), the active Na$^+$ transfer from inside the tubular cell into the blood increases without limits with increasing flow of glomerular filtrate: Cl$^-$ *and water passively follow from lumen to blood through the large surface provided by the brushed cells border towards the lumen* (Fig. 4.16).

The osmotic pressure inside and outside the tubule is the same (0.3 osmol/l); this poses the problem of how a large flow of fluid (7/8 of glomerular filtrate) takes place between two compartments with the same osmotic pressure. The large flow of liquid absorption is not explained by a difference in solute concentration and consequently by a difference in osmotic pressure between the two compartments. A probable mechanism is explained in Fig. 4.16. Between the cells of the proximal tubule exist *intercellular spaces where Na$^+$ is actively excreted as described above* (Fig. 4.16) *creating a gradient of concentration between lumen and intercellular spaces*, water follows and the intercellular spaces inflate exerting an hydraulic pressure which pushes the isotonic liquid through the basement membrane into the blood capillary.

4.6 Loop of Henle

Now that we have seen how kidney maintains constant the relative concentration of the different substances in the blood, we must understand two other functions of kidney: the regulation of the osmotic pressure and of the pH. If the concentration is 1 osmol/l, the osmotic pressure is $22.4 \times 760 = 17,024$ mmHg (1 osmolarity = 6.02×10^{23} molecules/liter of water corresponding to 22.4 atmospheres). The urine can be usually concentrated up to 4 times, from the physiological 0.3–1.2 Osm, from ~5000 to ~ 0,000 mmHg: an increment of 15,000 mmHg i.e. ~20 Atmospheres! This implies pushing the piston of Fig. 4.3 with an exceedingly high force. Nature solved the problem by means of the *loop of Henle*. When urine is concentrating the osmotic pressure increases *equally* inside the tubule *and* outside of it in the interstitium from *cortex*, where the glomeruli are, to *medulla* attained by the loops of Henle (Fig. 4.1, right). In this way no appreciable gradient of osmotic pressure exists between inside of the tubule of the loop of Henle and the adjacent tissues where osmotic work is done. It is the *collecting tubule* running from cortex to medulla, that *passively* equilibrates its inside with the increased osmolarity of the surrounding tissues delivering to the ureter the concentrated urine (Fig. 4.17). Obviously urine can be diluted as well with the same mechanism.

Figure 4.17, shows how this is done in detail. *The descending limb of the loop of Henle is passively permeable to water and solutes*; on the contrary the ascending limb is *impermeable to water* and is the site where work is done by *active secretion of Na$^+$ (and Cl$^-$?). The transport of water and solutes (Na$^+$) from the ascending tubule to the interstitium is facilitated by the hormone aldosterone.* Blood within *vasa recta* runs between the two tubules of the loop of Henle with a flow direction *opposite* that within the loop; in the example of Fig. 4.17, their osmolarity is the physiological one (0.3 Osm) when entering the loop and 0.35 when leaving it, indicating a recovery of some excreted solutes. On the contrary at the end of loop of Henle the osmolarity

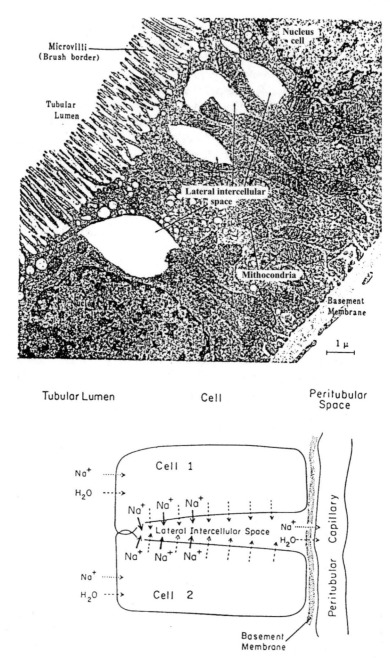

Fig. 4.16 Postulated mechanism for isosmotic fluid transport by epithelial membranes. At the top is an electron micrograph of proximal tubular epithelium from a rat; it serves to orient the model at the bottom. Solid arrows = active Na^+ transport; dotted arrows = passive Na^+ transport and interrupted arrows = passive water movement (from Valtin H *Renal function*, Little Brown and Company, 1973)

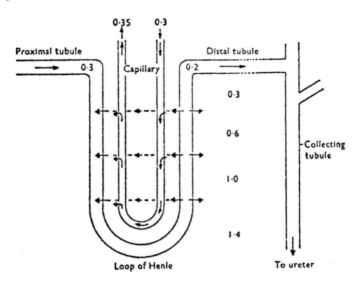

Fig. 4.17 The countercurrent multiplier system for the formation of concentrated urine. It is shown a case where urine is highly concentrated from the 0.3 osmolarity of plasma to a maximum of 1.4 Osm in the urine. The arrows in the *ascending* limb of the loop of Henle indicate *active excretion* of salt (NaCl) not followed by water. The excreted salt diffuses horizontally into the descending limb of the loop of Henle both through the interstitium (interrupted horizontal lines) and by a capillary blood with flow opposite that in the loop of Henle (continuous lines inside the capillary). In this way osmotic pressure increases from cortex to medulla of the kidney and in the permeable collecting tubule drawn on the right of the Figure (from Green JH *An Introduction to Human Physiology*, Oxford University Press, London, 1968)

(0.2) is smaller than at the beginning (0.3) indicating an *exit of solutes not followed by water.* The active excretion of salt (NaCl) by the ascending limb of the loop of Henle implies *work* to be done, but *not* against an osmotic pressure gradient because the solute actively excreted by the bottom of the ascending limb is absorbed by the adjacent descending limb and returns back into the ascending limb with the consequence that osmotic pressure is the same inside the loop of Henle and in the tissues outside of it. This is achieved through a *countercurrent transverse transport* of solute which takes place both directly from ascending to the nearby descending limb of loop of Henle *and* through the *vasa recta* as indicated by the arrows in Fig. 4.17. Since NaCl begins to be excreted and 'trapped' at the bottom of the ascending limb, its concentration is maximal at the bottom and progressively decreases from medulla to cortex. Osmolarity will increase from cortex to medulla of kidney up to very high values.

To summarize, the functions of *vasa recta are*: (i) contribute to the *transversal* transport of solutes from the ascending to the descending tubules of the loop of Henle as described above; (ii) take away the excess of water and salt; in this sense the *vasa recta* partially undo what the Loop of Henle does by recovering some of the salt and water from the interstitium. Suppose that active excretion would stop: vasa recta will take away the solute tending to make the osmotic pressure of the whole interstitium 0.3 Osm; (iii) in stationary conditions, due to the active excretion of salt by the ascending limb, the descending limb of the loop of Henle is surrounded by a progressively concentrated interstitium; it follows that water will tend to flow from inside the descending limb into the interstitium swelling the kidney tissue. This is prevented by the vasa recta, which 'take away' the excess of water and solutes.

As shown in Fig. 4.17 the osmolarity of the fluid contained in the *distal convoluted tubule* is low; subsequently however the *collecting tubule*, passes from cortex to medulla through an interstitium where the solute concentration progressively increases with the mechanism described above (Figs. 4.17 and 4.18). It follows that if the wall of the collecting tubule are permeable to water, the urine concentration *passively* increases by equilibrating with surrounding tissue thanks to the osmotic work done by the loop of Henle against a nil osmotic pressure gradient. *The porosity of the collecting tubule walls is increased by the antidiuretic hormone ADH* (also called *vasopressin* because it induces contraction of the smooth cells in blood vessels), which is produced by the nervous cells of the supraoptic nucleus in hypothalamus, with their axons releasing ADH into the posterior pituitary gland. Main trigger to ADH release into the blood is a high osmotic pressure of blood (due to high concentration of salt, proteins etc.). The action of ADH is to widen all the porosity allowing exit of water not followed by solutes; in this way urine is concentrated and water is retained in our organism when needed. As mentioned above, glomerular filtration is ~120 ml/min (Fig. 4.5). Of these 120 ml/min, 7/8, i.e. 105 ml/min, are *always* reabsorbed by the proximal convoluted tubule (*obligatory reabsorption*), the other 15 ml/min continues along the nephron and are reabsorbed as described above according to the physiological conditions (*optional reabsorption*). *In absence of the ADH* or of its function on kidney (as it may occur in some pathological conditions), 15 ml/min corresponding to ~20 l of urine are eliminated in one day, these patients are very thirsty; this condition is called *diabetes insipidus*. Also in case of an excess of glucose in the blood (diabetes) a large volume of urine is eliminated because all of the glucose transporters are saturated and high glucose concentration prevents water absorption in the collecting tubule (osmotic diuresis). In the old days. doctors were tasting the urine of patients to decide if the excessive diuresis was due to an excess of glucose "sweet" (osmotic diuresis due to an excess of glucose) or was 'insipidus' (prevented reabsorption of water due to inefficient ADH action).

The daily volume of urine is reduced to an *obligatory minimum* when (as lost in a desert or at sea) no water is available to drink. Let's consider an example of this condition. Concentration C is quantity Q of solutes eliminated with urine divided by the volume of urine V eliminated in the same time, i.e. $C = \dot{Q}/\dot{V}$; i.e. $\dot{V} = \dot{Q}/C$. Reasonable values in normal condition is $Q = 0.96$ Osm/day, $\dot{V} = 1.2$ l/day and $C = 0.8$ Osm/l. Now let's go in the desert without water to drink; if the moles to eliminate remain the same the flow of urine would be $\dot{V} = 0.96/1.4 = 0.69$ l/day

Fig. 4.18 Mechanism of
concentration of urine. The
collecting tubule passes
through tissues spaces
having an increasing
osmolarity. As a result water
is absorbed and the urine
becomes more concentrated
up to a maximum of
1.4 Osm/l (from Green JH
*An Introduction to Human
Physiology*, Oxford
University Press, London,
1968)

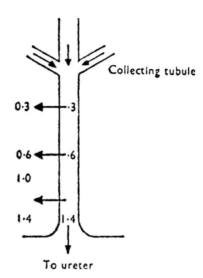

(note that as shown in Figs. 4.17 and 4.18, the maximal concentration the kidney can attain is 1.4 Osm/l). Fortunately in the desert we will no eat with the consequence that the osmoles to be eliminated will decrease say to 0.8 Osm with the consequence that the minimal obligatory volume will decrease to $\dot{V} = 0.8/1.4 = 0.57$ l/day. When fasting we 'eat' our fat reserve, but if we had at disposal some sugar we will 'burn' it: $C_6H_{12}O_6 + 6O_2 = 6CO_2 + 6H_2O$, i.e. *water* is produced with the consequence that 'fasting' with glucose the condition will be $\dot{V} = 0.2/1.4 = 0.14$ l/day, i.e. the minimum volume (obligatory volume) necessary to excrete the waste products will decrease. In conclusion to decrease the volume of urine to a minimum, since $\dot{V} = \dot{Q}/C$ we must decrease the amount of waste Q given that the maximum concentration C attained by the kidney is 1.4 Osm/l.

Figure 4.19 summarizes some of the mechanisms hitherto described. A decrease in blood volume and pressure is sensed by the juxtaglomerular apparatus of the *macula densa* consisting in cells of the distal tubule positioned at the angle between afferent and efferent arterioles. The juxtaglomerular cells secrete *renin*, which combines with angiotensinogen to produce *angiotensin*, a peptide hormone that stimulates the release of *aldosterone* from the cortex of the suprarenal glands. *Aldosterone promotes the transfer of solutes from the ascending tubule of the loop of Henle causing an increased osmotic pressure in the interstitium* (Fig. 4.17). The high osmotic pressure of blood due to the decreased blood volume triggers release of the *antidiuretic hormone* ADH, which is produced by the nervous cells of the supraoptic nucleus in hypothalamus. As described above *ADH increases porosity of all membranes allowing water and solutes to be reabsorbed by the interstitium into blood vessels thus increasing blood volume and pressure.*

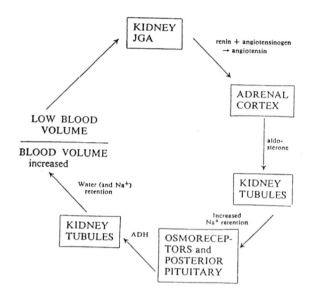

Fig. 4.19 Schema of the iuxtaglomerulus apparatus (JGA) reassuming the feedback system allowing regulation of blood volume and pressure with the mechanisms described in the text (from Green JH *An Introduction to Human Physiology*, Oxford University Press, London, 1968)

4.7   Distal Convoluted Tubule

4.7.1   Cycle of Urea and Osmotic Diuresis

As mentioned above *urea* is absorbed 'passively' (Fig. 4.14) thanks to the concentration condition of the glomerular filtrate attained by the active excretion of other substances. At the end of the proximal convoluted tubule 7/8 of filtrate are reabsorbed and half of the filtered urea is reabsorbed (50% in Fig. 4.20). However at the *beginning* of the distal convoluted tubule the amount of urea is 10% *greater* than that filtered (Fig. 4.20) indicating that urea enters within the tubules of the loop of Henle coming out of the *collecting tubule*.

In conclusion of the total amount of urea filtered at the level of the Bowman capsule (100 in Fig. 4.20), 60% are reabsorbed and only 40% are eliminated with the urine. This is a relevant fraction of the urine osmolarity (total: 2000 mOsm/kg, urea 800 mOsm/kg).

Figure 4.21 shows how this happens. Consider the case of diabetes, with a very high concentration of glucose in the blood; at the level of the collecting tubule the concentration of glucose will increase to such an extent to limit the transfer of water to the renal interstitium resulting in a loss of water with the urine; this is called *osmotic diuresis* of glucose. This would also be the case of the *urea*, which is an important constituent of the urine requiring the highest osmotic work to be eliminated (Table 4.1). As shown in Fig. 4.20, the amount of urea in the *distal tubule* is 10% greater than in the initial filtrate. The high concentration of urea in the *collecting tubule* would cause an osmotic diuresis of urea preventing further transfer of water from the collecting tubule into the interstitium. This is prevented by a different effect

Fig. 4.20 Renal handling of urea at normal rates of urine flow. The numbers within the lumen denote the percentage of the filtered amount of urea that flows at the various sites. The arrows indicate the medullary recycling of urea (adapted from Lassiter W, Gottschalk CW and Mylle M *Am. J. Physiol.* 200: 1139, 1961)

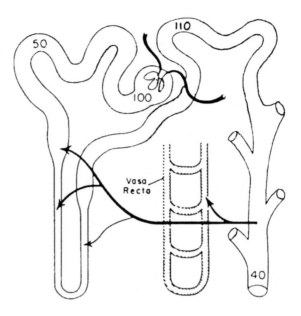

of ADH on the cortical and medullar portions of the collecting tubule as described in Fig. 4.21: contrary to the cortical portions, the medullary portion of the collecting tubule is made permeable by ADH to urea, which exits with water from the tubule into interstitium preventing an excessive increase of its concentration in the collecting tubule which would cause an osmotic diuresis of urea. This mechanism implies a greater concentration of both NaCl and *urea* in the papillary tissue of kidney.

4.8 pH Regulation by Kidney

Three mechanisms, involving *bicarbonate* ions, *phosphates* and *ammonia*. As we have seen: pH $= 6.1 + \log ([HCO_3^-]/\alpha pCO_2)$. At high altitude the glomus cells located in the carotid bodies and aortic bodies, sense hypoxia and cause hyperventilation decreasing pCO_2; the consequent increase in pH is checked by kidney through an increased elimination of $[HCO_3^-]$. The mechanism is explained in Fig. 4.22.

CO_2 diffuses freely in out the cell as indicated by the horizontal arrows in Fig. 4.22. *If pCO_2 is decreased at high altitude* in our lungs and consequently in the peritubular fluid (due to hyperventilation), less $H_2CO_3 \leftrightharpoons H^+ + HCO_3^-$ will be formed inside the tubular cell, less H^+ will be *actively excreted* (as indicated by the circular arrow) into the tubule to combine with HCO_3^- and more of the *fast* flowing filtered bicarbonate ions HCO_3^- will be lost in the urine. As a consequence of these processes, accelerated by the enzyme carbonic anhydrase C.A. (present in the proximal tubule only to cope with the high velocity of flow) and by the large surface offered by the brushed interior

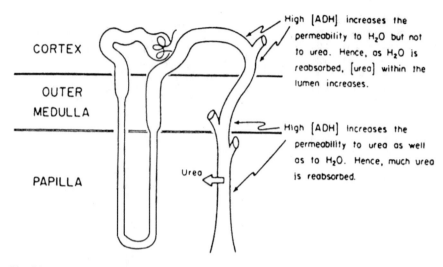

Fig. 4.21 Mechanism whereby urea increases the osmolality of the papillary interstitium and hence aids the conservation of water (from Valtin H *Renal function*, Little Brown and Company, 1973)

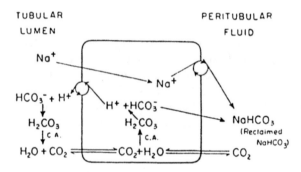

Fig. 4.22 Mechanism for the reabsorption of filtered HCO_3^-. C.A. stands for the enzyme *carbonic anhydrase*, which catalyzes in both directions the reaction $CO_2 + H_2O \rightleftharpoons H_2CO_3$. Most of this mechanism (80–90%) takes place in the proximal tubule (from Valtin H *Renal function*, Little Brown and Company, 1973)

of the proximal tubule (Fig. 4.16), the pH is maintained constant because the ratio $[HCO_3^-]/\alpha pCO_2$ is maintained constant due to a decrease of both the denominator (initial cause) and the numerator (subsequent remedy). *The mechanism is reversed in hypercapnia*, as e.g. in the *respiratory insufficiency* when *respiratory acidosis* takes place. In this case the increased pCO_2 'pushes' H^+ out decreasing the loss of HCO_3^-, which is retained tending to maintain constant the ratio $[HCO_3^-]/\alpha pCO_2$ and the pH of the organism. Note that the *bicarbonate* ions mechanism, just described, does not imply a net excretion of H^+, whereas the phosphate and ammonia mechanisms that will be described below do imply a net excretion of H^+ in the urine. *The bicarbonate ions mechanism just balances the amount of HCO_3^- to be excreted or conserved* (the

Fig. 4.23 Mechanism whereby titratable acid (T.A.) is created and newly formed HCO_3^- is added to the blood along with reabsorbed Na^+ (from Valtin H *Renal function*, Little Brown and Company, 1973)

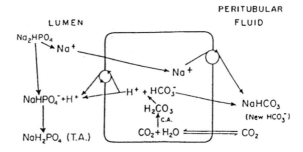

alkaline reserve). Since in the proximal tubule the flow is 80–90% greater than in the distal tubule and in the collecting duct, the largest amount of HCO_3^- absorption takes place in the proximal tubule with relatively small variation of pH in the primitive urine. In the distal portion of the nephron the HCO_3^- absorption is much smaller, but the consequent pH difference between inside and outside of the tubules becomes larger: the H^+ pump (circle in the left side of Fig. 4.22) must work against a greater H^+ concentration within the tubule, the *extensive* work done in the proximal tubule is substituted with an *intensive* work in the distal tubule. In order to decrease the H^+ concentration against which the pump must work, the H^+ concentration within the lumen is contained by *buffer systems*. Figure 4.23 shows the *buffer system of phosphates*: pH $= 6.8 + \log ([HPO_4^{--}]/[H_2PO_4^-])$.

Figure 4.23 shows that a sodium ion Na^+ initially equilibrating one of the two negative charges of HPO_4^{--} is substituted by a hydrogen ion H^+ actively excreted (round arrow) or passively diffused (straight arrow) depending on a greater or lower concentration of H^+ into the lumen. The origin of the excreted H^+ is carbonic acid formed within the cell (thanks to the carbonic anhydrase, C.A.) from the CO_2 diffused from the peritubular fluid. Note that differently from Fig. 4.22, CO_2 is used to provide *new* H^+ lost in the urine as titratable acid (T.A.), and newly formed HCO_3^-. In other words $[CO_2]$ is not the same in lumen and peritubular fluid as in Fig. 4.22: *CO_2 is lost by the organism to provide new H^+ to be excreted with the urine and new HCO_3^- to be adsorbed in the blood*. The pH of the urine can range from a maximum of 8.4 to a minimum of 4.4, i.e. $8.4 = 6.8 + \log ([HPO_4^{--}]/[H_2PO_4^-])$; $8.4 - 6.8 = 1.6 = \log 40$ and $4.4 - 6.8 = \log -2.4 \approx 1/250$. If the pH of urine is 4.4, since the pH of plasma is 7.4, the difference is 3, indicating a H^+ concentration 1000 times greater in urine than in plasma! This is the maximum acidity of urine. From what hitherto described it appears that 'extensive' work is done in the proximal portion of the nephron and 'intensive' work is done in its medullar portions. A third mechanism is activated in these conditions: the production of ammonia NH_3 by the kidney (Fig. 4.24).

As always, the CO_2 in the peritubular fluid diffuses within the renal cell and, thanks to the accelerating enzyme carbonic anhydrase (C.A. present inside the cell only), is hydrated to carbonic acid, which provides hydrogen ions to be excreted and bicarbonate ions to be adsorbed. The renal cell produces ammonia NH_3, which is very soluble into the membrane and pass in the tubule to combine with the excreted

Fig. 4.24 Mechanism for the renal excretion of NH$_4^+$(from Valtin H *Renal function*, Little Brown and Company, 1973)

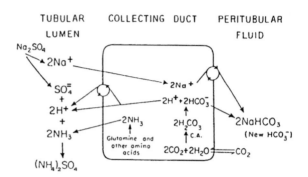

hydrogen ions to form ammonium ions NH$_4^+$(which are not soluble into the membrane, remain trapped in the tubule and eliminated with the urine as ammonium sulfate. This derives from sodium sulfate providing the two sodium ions adsorbed in the tubular fluid as sodium bicarbonate. *The transport of water and solutes (Na$^+$) from the ascending tubule into the interstitium, facilitated by the hormone aldosterone, takes place in exchange with active H$^+$ excretion* (Fig. 4.24). *Active excretion of H$^+$ is in competition with excretion of potassium ions K$^+$.* Ingestion of potassium salts decreases excretion of H$^+$ with the consequence that less HCO$_3^-$ is adsorbed and is eliminated with the urine causing osmotic diuresis by HCO$_3^-$. Acidosis has an opposite effect: more H$^+$ are excreted stopping the flow of Na$^+$ HCO$_3^-$, but the concentration of K$^+$ in the organism will increase. Vomiting and diarrhea cause a large loss of cells containing K$^+$: by competition more H$^+$ will be eliminated and alkalosis will ensue.

4.9   Water Balance of Our Organism

We are composed mainly by water (~60% of our body mass). As indicated in Fig. 4.25, the inlet of water takes place by drinking, but also by eating because most foods contain water; we have seen that burning glucose: C$_6$H$_{12}$O$_6$ + 6O$_2$ = 6CO$_2$ + 6H$_2$O.

Loss of water takes place mainly through urine and evaporation through the skin and the expired air. However particular attention must be given to the diarrhea in the newborn frequently resulting in dehydration.

4.9.1   Fluid Compartments of the Organism

Some are measured, some are calculated. Measurement is based on the usual equation: Volume = Quantity/Concentration, $V = Q/C$. The idea is to introduce a known

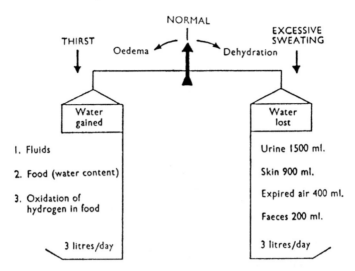

Fig. 4.25 Water balance of the body in normal resting condition (from Green JH *An Introduction to Human Physiology*, Oxford University Press, London, 1968)

quantity Q of substance and to measure its concentration C after mixing and before losses. The volume of *plasma* (~3 l) is measured after injection of *Evans Blue* (dye), which remains trapped in the blood vessels. The total volume of water in and out the cells is measured injecting a known quantity of *heavy water* either deuterium oxide (D_2O) or tritium oxide (T_2O): this volume in the adult is 45 l. The volume of *extracellular* water is measured using molecules (as radioactive sodium), which hardly enter the cells: 15 l in the adult. From these measured values one can calculate: the volume of intracellular water: $45 - 15 = 30$ l and the volume of tissue interstitial fluid $15 - 3 = 12$ l in the adult (Fig. 4.26).

Figure 4.27 shows that each water compartment is electrically neutral, i.e. the concentration of the negative ions equals that of positive ions. It also shows that the total ion concentration is greater in the intracellular fluid, intermediate in plasma and minimal in the tissue interstitial fluid, i.e. is greater where is greater the concentration of protein non diffusible ions; it follows that the osmotic pressure is greater in the compartments containing proteins not only due to their concentration, but also due to a greater concentration of diffusible ions. The reason for this difference is explained below.

4.9.2 Donnan Equilibrium

Figure 4.28 shows two compartments (1 and 2) divided by a membrane allowing transit of sodium ions (Na^+) and chloride ions (Cl^-) but not of protein ions (P^-). The probability p that a Na^+ ion on compartment 2 faces a pore of the membrane

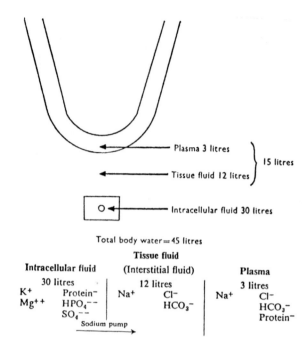

Fig. 4.26 Fluids compartments of the body with their principal ions content (from Green JH *An Introduction to Human Physiology*, Oxford University Press, London, 1968)

is proportional to its concentration in compartment 2, i.e. $p_1 = k_1 \, [Na^+]_2$. To *keep electrical neutrality* during the transit from compartment 2 to compartment 1 also a Cl^- ion must face the same pore at the same instant with a probability proportional to its concentration, i.e. $p_2 = k_2 \, [Cl^-]_2$. The probability that *both* ions meet in the same place at the same instant is proportional to the product of the two concentrations: $p = k \, [Na^+]_2 \times [Cl^-]_2$. The same must hold for the transit from compartment 1 to compartment 2 after Cl^- entered into it, i.e. $p = k \, [Na^+]_1 \times [Cl^-]_1$. At equilibrium the two probabilities are equal: $[Na^+]_2 \times [Cl^-]_2 = [Na^+]_1 \times [Cl^-]_1$. In Fig. 4.28: $(b - x)^2 = (a + x) \times x$. We can see this geometrically as the area of a square $(b - x)^2$ being equal to that of a rectangle $(a + x) \times x$: e.g. $2 \times 2 = 4 \times 1$, but this involves that $2 + 2 < 4 + 1$ because the sum of the sides of a rectangle is greater than the sum of the sides of a square having the same area; i.e. *the concentration of diffusible ions is greater in the compartment containing the non-diffusible ions* (Fig. 4.28, first consequence of Donnan equilibrium). This means that the osmotic pressure in the compartment containing proteins is greater than that in the compartment without proteins for two reasons: (i) the *oncotic pressure* of proteins, and (ii) the greater concentration of diffusible salt ions. For this reason a piston is drawn in Fig. 4.28 to prevent increase in volume of compartment 1. For the same reason gelatin swells if immersed in salt water.

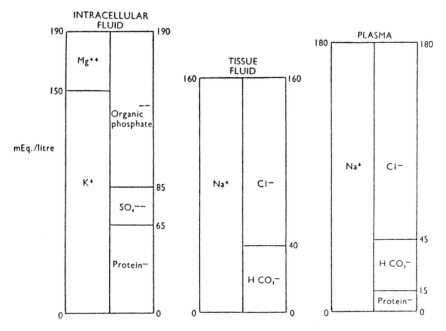

Fig. 4.27 Positive and negative ions in the water compartments of the body (from Green JH *An Introduction to Human Physiology*, Oxford University Press, London, 1968)

Fig. 4.28 Two compartments (1 and 2) divided by a membrane allowing passage of sodium ions (Na^+) and chloride ions (Cl^-) but not to protein ions (P^-); the gray piston on the left compartment impedes volume changes. The two compartments above indicate the starting condition, which evolves into the equilibrium condition as indicated below

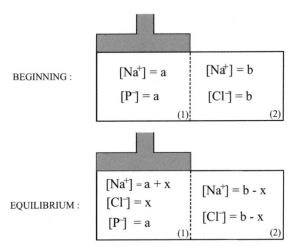

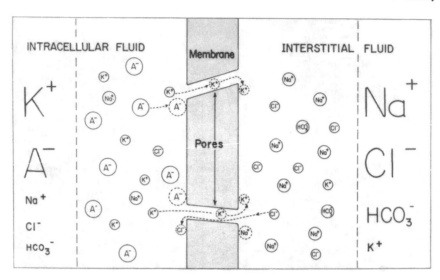

Fig. 4.29 Left: inside of a cell; positive potassium ions K⁺ and negative protein ions A⁻ (P⁻ in Fig. 4.28) are the greatest components, followed by sodium, chlorine and bicarbonate ions. Right: the positive sodium ions Na⁺ and Cl⁻ outside of the cell are the greatest components followed by bicarbonate ions and potassium ions. Middle: cell membrane is drawn with two pores. The interrupted lines through the pores indicate the transfer of positive ions from inside to outside the cell restrained by the attraction of negative ions inside the cell (from Ruch TC & Patton HD *Physiology and biophysics vol II*, W.B. Saunders Company, Philadelphia London Toronto, 1974)

As described above, at equilibrium: $[Na^+]_1 \times [Cl^-]_1 = [Na^+]_2 \times [Cl^-]_2$. It follows that $[Na^+]_1/[Na^+]_2 = [Cl^-]_2/[Cl^-]_1$, i.e. $(a+x)/(b-x) = (b-x)/x$. In our example: $4/2 = 2/1$. This means that, in absence of active transport, the concentration of the *positive* diffusible ions ($[Na^+]$ in our example) is greater in the compartment containing the negative non-diffusible ion ($[P^-]$). The contrary is true for the *negative* diffusible ions, which are in greater concentration where the concentration of the negative non-diffusible ions is less. This is the second consequence of the Donnan's equilibrium.

From Fig. 4.28 it appears that the 'bumping' of the *individual* Cl⁻ ions from right (2) to left (1) will be greater than that from left to right; the opposite is true for the Na⁺ ions. This condition results in a *non-neutral-electric-transfer* of the individual ions. Figure 4.29 shows what happens in a real cell: some K^+ ions, in greater concentration inside the cell, pass through the membrane pore, but you can imagine them attached to a 'spring' keeping them attached to the outer side of the membrane; this 'spring' is in fact the electrical attraction of the negative protein ion A^- on the inside surface of the membrane. Similarly the Cl⁻ ions, in greater concentration outside the cell pass through the pore, but remain attached to the inside surface of the membrane by the electrical attraction of the positive K⁺ and Na⁺ on the outside surface of the membrane (bottom of Fig. 4.29). In this way we can imagine the whole system as made up by three neutral systems: interior of the cell, tissue outside the cell and the membrane (internal and external surface).

The different concentrations c of ions between adjacent compartments (Figs. 4.28 and 4.29) implies osmotic work $W_o = nRT \ln c_2/c_1$ done to create their difference in kinetic energy tending to invade the less concentrated department (see equations in Sect. 4.2 and Fig. 4.3). At equilibrium, the osmotic work equals the electric work W_e arising from the different electrical potential between inside and outside of adjacent compartments. Electrical work is defined as: $W_e = n E_p F z$, where n is the number of moles transferred against the electrical potential E_p; F, the Faraday number, is the amount of electric charge carried by one mole of electrons (96,500 C/mole), and z is the valence, i.e. the number of electrons per mole. Therefore at equilibrium: $nRT \ln c_2/c_1 = n E_p F z$, i.e. $E_p = (RT/F z) \ln c_2/c_1$ which is the Nernst equation indicating the electrical potential difference created by the different ions concentration inside and outside the cell.

Printed in the United States
By Bookmasters